Mars One
The Ultimate Reality TV Show?

Erik Seedhouse

Mars One

The Ultimate Reality TV Show?

Published in association with
Praxis Publishing
Chichester, UK

Erik Seedhouse
Assistant Professor
Aviation Sciences Embry-Riddle Aeronautical University
Daytona Beach, Florida
USA

SPRINGER-PRAXIS BOOKS IN SPACE EXPLORATION

Springer Praxis Books
ISBN 978-3-319-44496-3   ISBN 978-3-319-44497-0   (eBook)
DOI 10.1007/978-3-319-44497-0

Library of Congress Control Number: 2016951723

Cover design: Jim Wilkie

Project Editor: Michael D. Shayler

Printed on acid-free paper

This Springer imprint is published by Springer Nature
The registered company is Springer International Publishing AG Switzerland
The registered company address is: Gewerbestrasse 11, 6330 Cham, Switzerland

Contents

Acknowledgements

In writing this book, the author has been fortunate to have had five reviewers who made such positive comments concerning the content of this publication. He is also grateful to Maury Solomon at Springer and to Clive Horwood and his team at Praxis for guiding this book through the publication process. The author also gratefully acknowledges all those who gave permission to use many of the images in this book. The author also expresses his deep appreciation to Mike Shayler, whose attention to detail and patience greatly facilitated the publication of this book and to Jim Wilkie for creating yet another striking cover. Thanks Jim! Thanks also to Senior Project Manager, Sarumathi Hemachandirane, for guiding this project through the editing process.

To SpaceX – the most likely to reach Mars first

About the Author

Figure P.0   Credit: Erik Seedhouse

Erik Seedhouse is a fully-trained commercial suborbital astronaut. After completing his first degree he joined the 2nd Battalion the Parachute Regiment. During his time in the 'Paras', Erik spent six months in Belize, where he was trained in the art of jungle warfare. Later, he spent several months learning the intricacies of desert warfare in Cyprus. He made more than 30 jumps from a Hercules C130 aircraft, performed more than 200 heli-copter abseils and fired more light anti-tank weapons than he cares to remember!

Upon returning to academia, the author embarked upon a Master's degree, which he supported by winning prize money in 100km running races. After placing third in the World 100km Championships in 1992, Erik turned to ultra-distance triathlon, winning the World Endurance Triathlon Championships in 1995 and 1996. For good measure he won the World Double Ironman Championships in 1995 and the infamous Decatriathlon, an event requiring competitors to swim 38km, cycle 1800km, and run 422km. Non-stop!

In 1996, Erik pursued his PhD at the German Space Agency's Institute for Space Medicine. While studying he found time to win Ultraman Hawai'i and the European Ultraman Championships, as well as completing Race Across America. Due to his suc-cess as the world's leading ultra-distance triathlete, Erik was featured in dozens of maga-zine and television interviews. In 1997 GQ magazine nominated him as the 'Fittest Man in the World'.

In 1999 Erik took a research job at Simon Fraser University. In 2005 the author worked as an astronaut training consultant for Bigelow Aerospace. Between 2008 and 2013 he served as Director of Canada's manned centrifuge and hypobaric operations. In 2009 he was one of the final 30 candidates in the Canadian Space Agency's Astronaut Recruitment Campaign. Erik has a dream job as an assistant professor at Embry-Riddle Aeronautical University in Daytona Beach, Florida. In his spare time he works as an astronaut instructor for Project PoSSUM, an occasional film consultant to Hollywood, a professional speaker, a triathlon coach and an author. 'Mars One' is his 25th book. When not enjoying the sun and rocket launches on Florida's Space Coast, he divides his time between his second home in Sandefjord and Waikoloa.

Acronyms

ALARA	As Low As Reasonably Achievable
ARS	Acute Radiation Syndrome
ARS	Atmosphere Revitalization System
ASV	Air Selector Valve
BEO	Beyond Earth Orbit
BFOQ	Bona Fide Occupational Qualification
BFR	Big Falcon Rocket
CDRA	Carbon Dioxide Removal Assembly
CF	Canadian Forces
CM	Command Module
COSPAR	Committee on Space Research
CSA	Canadian Space Agency
CSF	Cerebrospinal Fluid
DHMR	Dry Heat Microbial Reduction
DXA	Dual X-ray Absorptiometry
EAC	European Astronaut Center
ECLSS	Environmental Control Life Support System
EDL	Entry, Descent and Landing
ESA	European Space Agency
ESWL	Extracorporeal Shock Wave Lithotripsy
FAA	Federal Aviation Administration
GCR	Galactic Cosmic Rays
GI	Gastrointestinal
H2M	Humans to Mars
HEPA	High Efficiency Particulate Air (filters)
HTP	Hypersonic Transition Problem
ICE	Isolated Confined Environment
ICP	Intracranial Pressure
ICU	Intensive Care Unit

IOC	International Olympic Committee
ISRU	In-situ Resource Utilization
ISS	International Space Station
IST	Increment Specific Training
JSC	Johnson Space Center
LEO	Low Earth Orbit
LOC	Loss of Crew
LOM	Loss of Mission
LSAH	Longitudinal Survey of Astronaut Health
MAG	Maximum Absorbency Garment
MCT	Mars Colonial Transporter
MDM	Multiplexer/Demultiplexer
MIT	Massachusetts Institute of Technology
MOXIE	Mars Oxygen In-situ Resource Utilization Experiment
MTV	Mars Transfer Vehicle
NBL	Neutral Buoyancy Laboratory
NCRP	National Council on Radiation Protection
RAD	Radiation Assessment Detector
SLS	Space Launch System
SPE	Solar Particle Event
STP	Supersonic Transition Problem
TDS	Terminal Descent Sensor
TPS	Thermal Protection System
TRN	Terrain Relative Navigation
VIIP	Visually Induced Intracranial Pressure

Preface

"It's interesting as a thought experiment. I mean, Lansdorp doesn't have $6 bn, but NASA does and in terms of going only one way – hey, we're all on a one-way trip to somewhere! But I think it's utterly fantastical that you'll fund a Mars mission with a reality TV show."

Robert Zubrin.

Full disclosure here. I applied to Mars One in the spirit of 'nothing ventured, nothing gained'. I figured I would give the enterprise the benefit of the doubt and if it all fell apart then at least I would have material for a book, which is why you're reading this! So I have some insight into the Mars One mission and I also happen to have a number of space industry colleagues who went through the selection process, some of whom provided insight into the reasons why Mars One will never get off the ground. Also, as Editor-in-Chief of the *Handbook of Life Support Systems for Spacecraft and Extraterrestrial Habitats*, I had access to many world class experts who weighed in on some of the many, *many* reasons why the Mars One life support system cannot work – not for long at any rate.

"None of the vehicles exist. And you can't just go to SpaceMart and buy those things. People don't know what astronauts do. We don't just go to astronaut school, take a few classes and then ride a station somewhere. That's not how it works. It's simply not the process; it's almost sweetly naive to see it like that. We astronauts are in fact intrinsically involved in mission design. And in making the process safe. And it's really hard to get to Space Station and back. We've killed people, just trying to get up to space and back. Even very recently. It is extremely difficult to do that [get there and back].

"And the astronauts are as pivotal in making the process safe as anyone else. It's not simply a case of going to classes and learning which knobs to flick. That's not our part in the process at all. To be a shuttle commander, you already have to be one of the top test pilots in the world. Just to get selected – to get a start in the process. And we don't choose those people just because we want qualifications. We need those skills to safely do it. And even still, we lost two shuttles."

Chris Hadfield, Canadian Space Agency astronaut.
Perhaps the world's best known astronaut and someone who happens to know

Figure P.1   Credit: Mars One

more than most about spending time in space.

Mars One is a bold and very popular initiative; plugging it into Google generates 136 million hits! It is also an initiative mired in controversy. Their astronaut selection process required only a self-secured medical with the candidate's doctor. No mental health exam. No psychological or psychometric testing. Not even a criminal background check! In fact, the selection process to date more closely resembles the casting process for a terrestrial reality TV show than a serious astronaut selection. And on the subject of reality TV, Mars One's deal with Endemol Productions has fallen through. Its 6 billion dollar price-tag is fantasy because it doesn't consider resupply, while Mars One's assertions that it can attract media revenue at the same level as the Olympic Games are optimistic at best. Then there's Mars One's available finances. These are listed at just $700,000 (!), which is why the organization has suspended its robotic mission. And then there's the MIT life support study that predicts the first colonist will die after 68 days. Ask *any* professional astronaut and they will say the same thing: the technology required to get a human being to Mars and keep them alive on the surface does not exist – certainly not on a budget of $6 billion at any rate. But according to Mars One, these astronauts and highly credentialed people at MIT are wrong! So what is the real story? Rip-off? Hoax? Scam? Does Mars One have any chance of succeeding? You will find some of the answers in this book.

I've written more than 20 books and I have never had cause to include a warning before, but the content of this book may be disturbing to some, especially the 'Mars in a decade' crowd. The subject matter of this book may significantly disrupt your perception and belief as to whether a manned Mars mission is possible and will almost certainly shake your worldview of the viability of Mars One. This is not intentional: it is just that sometimes the truth hurts. So, if you want to continue to live in a fantasy world in which manned Mars missions are achievable within a decade on a budget of $6 billion and in which humans arrive on Mars unscathed by deep space radiation please stop reading right now. And, if you are a Mars One disciple who truly swallows the make-believe Harry Potter world where this can happen, perhaps you can save a couple of hours of your time by giving this book to someone else. But, if you are capable of dealing with the facts, however uncomfortable these may be, and if you feel there is something not quite right with the Mars One delusion, then this book is for you.

To begin with, let's give credit where credit is due and say right here and now that Bas Lansdorp did not invent the concept of a Mars One mission. Before Bas, there was Dirk

Schulze-Makuch of Washington State University. Together with his colleague, Paul Davies, Dr. Schulze-Makuch dreamt up a one-way Mars trip as a way of making an interplanetary journey more affordable: by as much as 80% in fact.

> "Really, this isn't a joy ride. You have to understand that the motivation for doing this is to not only open up a human presence on another planet, but to provide the opportunity to do some fantastic, groundbreaking science."
>
> *Dr. Paul Davies, Washington State University*

Now that we're clear on that, let's move on to a snapshot of the Mars One endeavor, such as it is. It all began with a website, complete with substandard CGI video. The organization bandied about a few letters from aerospace companies and the odd resume of people with little or no understanding of long duration spaceflight. They called a press conference to announce they were taking applications from anyone who wanted to go to Mars and never return. Nothing about the mode of transport. Nothing about the mode of landing, or the life support system, or radiation protection, or... anything really. But it was a reality television show so the media – predictable as ever – lapped it up. In fact they jumped all over it. A few clicks on Google and the hacks had their story. Never mind that most of the stories were full of holes. A one-way trip to Mars? What crackpots would sign up for something like this? And is it even viable anyway? Well, it obviously must be, because Mars One rolled out a Nobel Prize winner to say he liked the idea, even though this particular Nobel Prize winner's area of expertise had nothing to do with getting to Mars. But Nobel Prize winners are smart, so if this guy says it's doable it must be doable. The application process? Simple. Anyone can apply. After all, on a one-way trip to Mars what possible need would there be for a doctor or an engineer or a pilot? A baker you say? Sign up! Thousands did. To be fair, not all the applicants were astronaut no-hopers; there were some serious applications in the mix. But as the months dragged on, the reality, feasibility and fairy tale of Mars One began to unravel; hence this book. Still reading? Great. Depending on your perception, Mars One lies somewhere between an illusion perpetrated by one of the great con artists of all time, a suicide mission or just a crackpot idea. Which brings us to the wake.

In October 2015, I was lucky enough to be invited to give a talk at the Shackleton Museum Autumn School in Athy, Ireland. My talk was on the parallels between exploration in the 'Heroic Age of Antarctic Exploration' and future Mars exploration. At some point in the lecture, I described the Mars One mission architecture. I explained that while such a plan may seem extreme to many, this mindset was much more accepted back in the days of Shackleton and Amundsen, because 100 years ago the crews knew full well there was a good chance they were never coming back. After the lecture, a lady came up to me to say how much she enjoyed my talk and went on to suggest that perhaps the Mars One crew's families should hold a wake in advance of the mission, given that the outcome will be a foregone conclusion.

For those of you who have had the good fortune to visit Ireland, you will be familiar with the 'craic'. 'How's the craic?' is a greeting used by just about everyone in Ireland, and is a custom as common as St Patrick's Day or enjoying a pint of Guinness. More than most, the Irish have a rich history when it comes to customs and traditional rites. Especially when it comes to funerals, or 'wakes', as the Irish refer to their way of celebrating the death of a loved one. And 'celebrate' is the right word, because a wake is no place for tears. It is more a way to celebrate the life of a loved one in a way unique to the Irish custom: with plenty of food and drinks! Why the talk of wakes in a book dedicated to spaceflight? Well, if – and it's a mighty, *mighty* big 'if' – this Mars One boondoggle is ever realized, then the mission planners could do worse than plan a wake in celebration of the very-soon-to-be-deceased crew.

The pre-launch wake will serve an important function, because it will not only mark the departure of the crew from their home for the very last time, but also provide some measure of comfort to their family members, relatives and friends, who will have an opportunity to say their final goodbyes to their nearest and dearest. Mars One take note. Now back to Mars.

With the Curiosity Rover (Figure P2) going about its business and the success of the *'The Martian'*, it's not surprising that hardly a day goes by without some mention of manned missions to Mars, whether it be in the form of press releases, YouTube videos, NASA's marketing campaign – or those Mars One folks. Now, the blogosphere is overflowing with the naysayers and promoters of this self-styled suicide mission, but why does it attract so much press? Well, for the 'Mars in a decade' crowd it represents a fast track to Mars, because this sort of a mission can be done on a budget. Ask any engineer and they will tell you that it will cost ten times as much to bring astronauts back from Mars as it will just to get them there in the first place. Now you may be thinking that these sorts of missions should be left up to government agencies, but anyone who thinks that should take a step back and consider what plans NASA has to get humans to Mars.

You may have read about the Space Launch System (SLS), the Orion crew capsule and wonderful cutting edge space technologies, and how these will eventually enable astronauts to travel to Mars. You may also have read about the critical work being done

Figure P.2 Credit: NASA

to ensure that humans will land on Mars sooner rather than later and, after reading all of this, you may be persuaded that the venerable agency has a plan to get us to the Red Planet. After all, those pretty graphics showing mission architectures must have come about after an awful lot of planning by some very clever people, right? Well, yes, but how NASA is organized and how it operates isn't as streamlined as some people may think. For example, you may think that NASA Headquarters tells the rest of the NASA Centers what to do and those centers do exactly that, but this not how the agency works. In short, form does not follow function. And all those directorates: the Science Mission Directorate (SMD); the Human Exploration and Operations Directorate (HEOMD); the Space Technology Mission Directorate (STMD). Take your pick. And all those boards and councils and working groups? Well, you would think by looking at all those teams that NASA has this 'Mission to Mars' business figured out, especially when you see all the neat graphics. The problem is that while all these groups are supposed to coordinate a plan to get astronauts to Mars, it can't be done, because there just isn't enough money and no strategic master plan. Don't blame NASA for this, because it isn't NASA's fault. The agency does an absolutely incredible job with a budget that continues to be slashed in real terms, but there is only so much NASA can do with the pittance it is given.

A case in point: in October 2015, the agency released *Journey to Mars*, an upbeat 36-page report that described how NASA will send its astronauts to Mars. In the document were plans for asteroid capture, a deep-space laboratory and all sorts of new technologies that would be developed to help astronauts live and work on the Red Planet. The problem with this opus was that there were two key elements missing: dollars and deadlines. And without those, nobody is going anywhere. The problem? Political viscosity and successive Administrations – the Obama one in this case – that cut hundreds of millions of dollars from NASA's Mars programs. In fact, one of the reasons the Mars program and others have been kept alive is because Congress has passed budgets that have exceeded the President's budget request. But keeping a program on life support and actually making progress are two different animals, and if the United States wants to avoid more delays then the funding issue has to be solved. For quite some time, NASA has never known how much funding it will be receiving in any one year, and that makes it nigh on impossible to make any long-term plans, never mind developing all the new technologies needed for such a venture. To have any chance of putting government-trained astronauts on the Red Planet, NASA will need a big increase in funding *and* a coordinated strategy. Without those two factors, boots on the ground anytime soon is pure fantasy. Let's take an example. Back in the days when the International Space Station (ISS) was being planned, there was a proposal to fly a centrifuge called the Centrifuge Accommodation Module (CAM). You can see an image of this in Figure P.3. Needless to say, the CAM never flew. Instead, it is now gathering rust in a parking lot somewhere in Tsukuba, Japan. Which is a shame, because if the CAM had flown we would know a hell of a lot more about artificial gravity than we do now. But, because of fiscal viscosity, bad planning and the lack of a coordinated long-term strategy, we are almost as much in the dark about this Mars-enabling technology as we were 20 years ago.

And data on artificial gravity is just one of many, *many* data sets we badly need to acquire before a serious mission to Mars can be mounted. What about multi-year missions on the ISS? I keep a keen eye on the goings on in the world of manned spaceflight but I don't seem to remember a single NASA tweet about two-year missions. Or even one lasting 18 months for that matter. Yet a trip to Mars will be a multi-year mission, so how on Earth does NASA plan on gaining those invaluable long duration data-points? Especially now that the ISS is slated for decommissioning in 2024. Hmmm. The lack of a coordinated plan pretty much answers *that* question as well. Another check item on the manned Mars mission list is the sample return mission. NASA pulled the plug on this years ago and there is no definitive plan to realize this mission anytime soon. This, and so many, many other enabling missions and technologies are firmly slotted in the 'TBD' and/or 'fingers crossed' pile, thanks to the slow, uncoordinated climate that is pervasive nowadays. Another example? The Low-Density Supersonic Decelerator (LDSD). As all of you Mars fans know, one of the biggest challenges to any Mars mission, whether you plan on returning or not, is actually landing, and this is where the LDSD might come in handy.

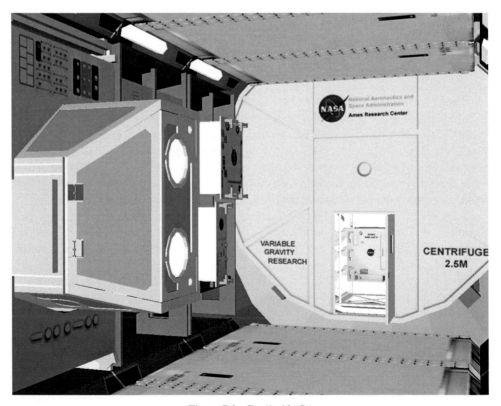

Figure P.3   Credit: NASA

NASA has tested the LDSD, and accumulated a fair amount of data until the parachute failed. The agency may do more testing or it may not. It depends, yet again, on funding and a coordinated plan. So, no artificial gravity, no LDSD and no multi-year missions. Let's face it, this 'manned mission to Mars' business is going nowhere fast. Sure, there are lots of graphics with arrows and flow charts with dates, but these do not constitute a viable plan to land humans on Mars in the 2030s. The 2130s perhaps. Pessimism? No, just a dose of reality. NASA has had a manned Mars mission on the drawing board for more than 40 years and all we have to show for those plans are some fancy graphics and charts. To many diehard Mars supporters, the 'Mars by the 2030s' mantra repeated by the NASA hierarchy is nothing more than a sales campaign and hot air, hence the support for other options – however misguided and far-fetched – such as Mars One. And even if there was a concerted effort starting today, we all know what can happen to best laid plans. Remember President Bush's Vision for Exploration (VSE)? It was announced in 2004 and it called for a return to the Moon by 2020 (Figure P.4). Well, since VSE was announced, the clock has been ticking steadily and NASA is nowhere near returning to the Moon. There is an outside chance a crew may fly a mission around the Moon sometime around 2023 but that speculative possibility hardly keeps the manned mission to Mars plan alive. The point is that NASA has been 20 years away from planning to send humans to Mars for the best part of half a century. Small wonder then, that a reality TV show called Mars One gets so much coverage: Mars supporters are just so fed up with listening to broken promises that they just want humans – any humans as it turns out – to get to Mars. Unorthodox? Certainly. Daring? Most definitely. Far-fetched and wildly ambitious? You betcha. Doomed to failure? More than likely. But for all the controversy surrounding this extravagant suicide mission, Mars One has captivated imaginations around the world by stating it will have humans on Mars by 2023. Or 2025. Or maybe 2027.

Mock them all you like, but you have to admire the bold idea of having everyday people pay money to sign up for the chance of being selected as an astronaut for a Mars mission. Harebrained? Absolutely. Feasible? No chance. We'll get into the details later in the book, but here are a few snapshots of why Mars One is doomed to fail. Let's begin with technology. In short there is a dearth of the technology needed to keep humans alive for months on end. In the world of manned spaceflight, this is known as life support. Now, if you listen to the Mars One debates and watch the interviews with the Mars One contestants, they will say all this life support technology exists. And to support this claim, the Mars One crowd uses the ISS as evidence, saying that these life support systems work just fine up there. Well, yes they do, but these systems won't work on Mars and here's why. On the ISS, the systems are replaced on a routine basis. That's the bonus of orbiting just 250 miles above the Earth: if something needs to be repaired, one of the supply vehicles can bring up the spare parts on the next resupply flight. Not so on a spacecraft en route to Mars. And on ISS, if something goes terribly wrong, the astronauts can simply pack up and head home in one of the lifeboat Soyuz craft. But if things go south on a spacecraft heading for the Red Planet? Well, let's just say it will be an unfavorable outcome for all concerned. Let's extend this logistical resupply issue a little. One thing NASA has learned over the years is that stuff breaks down. A lot. Which means resupply runs. Lots of them. In fact, the ISS gets resupplied every three months or so. But a Mars mission? We're talking about a long stretch of 26 months between resupply runs. No return vehicles for the Mars One

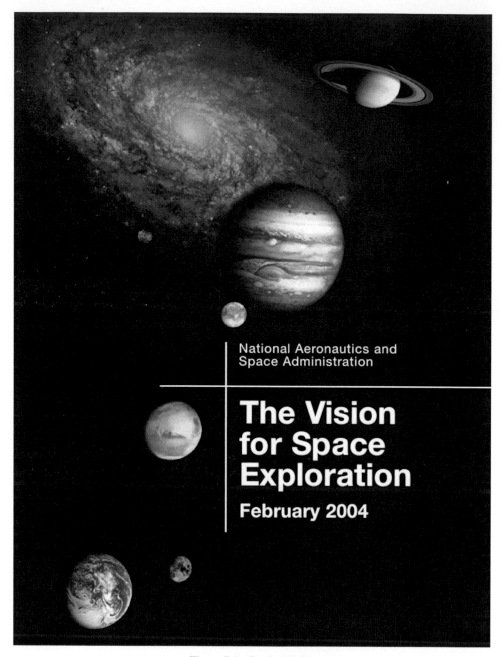

Figure P.4   Credit: NASA

astronauts, remember. Now, researchers have crunched the numbers and have reckoned that the Mars One crew will have to take along an awful lot of spare parts: based on the experience of astronauts living on the ISS over the years, three SpaceX Dragon vehicles will need to be stuffed with spare parts for every two crewmembers. And that's just to ensure a 50 percent chance of survival. That's bad enough, but the spare parts issue doesn't stop there, because the second crew will need to truck along its own spares and extra spares for the first crew. And so on and so on. Oh dear. Obviously such a mission plan is unsustainable. In-Situ Resource Utilization (ISRU) you say? Sure, but there is a *long* way to go before we have an ISRU system (Figure P.5) that is going to work reliably on Mars, as we'll find out later in this book.

But hasn't the Mars One team insisted that no new technology needs to be developed for this enterprise, such as it is? Yes, they have, which makes all these claims all the more surreal, because it would suggest that the Mars One team know something that space scientists and engineers don't: namely how to extract resources on the surface of Mars. Maybe they meant conceptually? Well, if they did then those in the Mars One team have a flimsy grasp of how conceptual understanding relates to actually doing stuff. After all, we have a pretty good idea of how fusion works, but nobody has actually gone ahead and built a fusion reactor because fusion is a tough nut to crack. As is ISRU as it turns out. And life support. And… well, you get the idea. Mars One defends itself against the naysayers on the subject of technology by insisting that the tech will be developed once they have investors. Catch-22 anyone? Let's face it: who in their right mind is going to sink money into a mission architecture when in reality there is no mission architecture? And there is no mission architecture because Mars One is still developing it. They don't know which vehicles will be ferrying their reality TV lemmings to the Red Planet and they haven't got a clue what sort of life support system they will be using. In short, the whole enterprise has no plan.

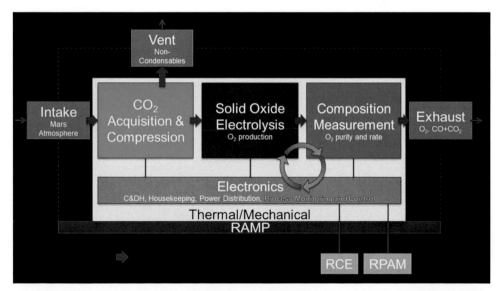

Figure P.5   Credit: NASA

OK, so enough about technology for the moment. What about some of the physiological mission-killers? Radiation, for instance. As long as there have been astronauts, we've known that space is a risky place to be. And one of the reasons it is so dangerous is radiation. For decades, we have been fed images and computer-generated videos of astronauts rambling around (Figure P.6) on the surface of Mars, when in reality they will be holed up in bunkers deep, *deep* underground. How little did we know. You see, radiation is the number one mission-killer in deep space and the take-home lesson from the results of all those radiation readings accumulated by probes sent to Mars[1] is this: take shielding – and plenty of it!

Now those of you who watched *The Martian* and believe (don't!) what they see in Hollywood movies may be wondering how Mark Watney survived. After all, NASA's most famous stranded astronaut spends most of the film wandering outside with little or no radiation protection to speak of, except for a space suit. In reality, Watney would have died of cancer, an outcome acknowledged by Andy Weir, the gifted author who wrote the book. Weir has acknowledged the lack of radiation protection in interviews, arguing that he had to apply some artistic license for the sake of moving the story forward. In the real world, the radiation measurements sent back by Curiosity indicate that after the outbound trip to Mars and 500 days on the surface, astronauts would receive a dose of one Sievert, which is about a five percent increased risk of cancer (NASA limits prescribe no higher than a three percent increased risk).

Now you may argue that this manned mission to Mars business is so risky that you just have to accept these hazards. Well, yes you do, but what happens to the crew when they start dying and losing their minds? You may be wondering why astronauts will lose their

Figure P.6   Credit: NASA

---

[1] The Mars Curiosity Rover calculated the average dose of a 180-day mission to Mars as 300 mSv, or about fifteen times the annual radiation limit for someone working in a nuclear power plant.

minds (some may argue that simply volunteering for Mars One would qualify, but that's another story), but the latest research shows that being exposed to cosmic rays for many months can result in brain damage and eventually dementia-like symptoms, such as memory deficits, confusion and loss of awareness. How do you care for an astronaut bumbling about a habitat unable to recognize his or her fellow crewmates and unable to perform the simplest of functions? It's a bad scenario at best. Mars One's answer? They don't have one and neither do any of the space agencies.

"We are very confident that our budgets, timelines and requirements are feasible."

Bas Lansdorp

You may be Bas, but the rest of us involved in the spaceflight industry are less than convinced. Your budget is a work of fantasy, your timeline is delusional at best and your requirements? Well, turn the pages and find out what the experts think about your requirements.

1

Mars One: The concept

Figure 1.0   Credit: Mars One

© Springer International Publishing Switzerland 2017
E. Seedhouse, *Mars One*, Springer Praxis Books, DOI 10.1007/978-3-319-44497-0_1

"This project seems to me to be the only way to fulfill dreams of mankind's expansion into space. It sounds like an amazingly fascinating experiment. Let's get started!"

Professor Dr. Gerard 't Hooft,
winner, 1999 Nobel Prize in Physics

The next giant leap for mankind? The next step of the voyage into the solar system? An inspiration for generations to come? Or a suicide mission dreamt up by a delusional madman, undertaken by a bunch of lunatic lemmings? Mars One has been called lots of things. Mostly unkind. But before we start looking at the finer details, let's examine the history of what is unarguably a unique venture, however misguided it is perceived by, and portrayed in, the popular press.

"Yes, but to travel to another planet, knowing you can never come back, you'd have to be pretty sad. Aniston sad."

Marge Simpson, The Simpsons, Season 27 Episode 16.
The Marge-ian Chronicles. *Original Air Date, March 13, 2016*

Before going into the details of the boondoggle that is Mars One, let's turn it over to the editors of the reference manual for this enterprise, a book titled *Mars One: Humanity's Next Great Adventure*. Big on entertainment value but skimpy on detail, this book was published in early 2016. Its editors define Mars One as:

"Mars One is a non-profit organization, based in the Netherlands and international in scope, whose goal is to establish a permanent human settlement on Mars. Why do this? Because it is the next giant leap forward for humankind; a stepping stone for the human race on its unyielding quest to explore the universe. Human settlement on Mars will aid our understanding of the origins of the solar system, the origins of life, and our place in the cosmos."

Sound grandiose doesn't it? The introduction to the Mars One book warbles on about the fantastic adventure that this Mars One trip will play out, and how awestruck the viewers at home will be as they watch these intrepid contestants take their first steps on the Red Planet. Keep reading the blurb and it's easy to be sucked into the fantasy – until you read the timeline. A timeline that has Mars One landing its first contestants in 2027! As anyone with even the flimsiest grasp of manned spaceflight will tell you, this goal is not just a fantasy but just plain MAD! Now the idea of a one-way mission has its merits, because such an expedition dramatically reduces consumables and fuel. But 2027? Get real! Astronauts haven't ventured beyond Earth orbit in more than 40 years and all of a sudden a reality television show is making the giant leap to Mars. There is just no way, is there? Well, no, but before we start dismantling what little mission architecture there is, we need to understand the background to this suicidal enterprise.

Mars One (Figure 1.1) was announced on May 31, 2012 in Amersfoort, The Netherlands. That was when the world was introduced to the concept of a one-way mission to the Red Planet (the idea has been around for a while but it was Mars One that pushed the idea into mainstream media), crewed by a motley team of reality TV contestants and financed by intellectual property rights, investments and donations. The plan: establish a human

Figure 1.1   Mars One logo. Credit: Mars One

settlement on the Red Planet in 2027 by first sending a crew of four, to be joined by further crews of four every two years. To cut down on mission complexity and costs, none of the Mars One contestants will return to Earth (for many of the detractors of the mission this is a good thing). Instead, they will live and work on Mars, mining resources from the soil, growing plants, and being filmed 24/7 for the worldwide media event that will become Mars One Reality TV. To show they were serious, Mars One took pains to point out that they had contacted various aerospace companies around the world – ILC Dover, MDA Information Systems, Paragon Space Development, SpaceX and Thales Alenia Space – and had received letters of interest from these suppliers. Here's a word from the man who dreamt up the idea, Bas Lansdorp:

> "Since its conceptualization, Mars One has evolved from a bold idea to an ambitious but feasible plan. Just about everyone we speak to is amazed by how realistic our plan is. The next step is introducing the project to the world and securing sponsors and investors. Human exploration of Mars will be the most exciting adventure mankind has embarked upon in decades. It will inspire a new generation of engineers, inventors, artists and scientists. It will create breakthroughs in recycling, life support and solar power systems. It will create a new generation of heroes – the first explorers to go to Mars will step straight into the history books. Finally, we expect it to capture an audience of millions, culminating in several billion online spectators when the first crew lands on Mars."

> *Bas Lansdorp, Mars One Chief.*

Figure 1.2 Bas Lansdorp. Credit: Joe Arrigo

BAS

Bas Lansdorp (Figure 1.2) is a Dutch entrepreneur who made his money by selling his shares in Ampyx Power, a wind energy company. His interest in sending humans to Mars began while studying at the University of Twente, although it was the business model that attracted him rather than overcoming the technological challenges. The trigger event for the idea of a one-way trip to the Red Planet came when Lansdorp saw revenue numbers published by the International Olympic Committee (IOC). Together with his co-founder, Arno Wielders, Lansdorp got in touch with Dutch media expert, Paul Römer, and pitched the idea of sending humans to Mars on a one-way trip. After some discussion, the three came up with the idea of a reality TV show. Anyone over the age of 18 could apply and the audience would democratically decide who the crew would be. *Big Brother* anyone (see sidebar)?

Big Brother on Mars?

Big Brother is a reality television show that takes its name from George Orwell's novel, *Nineteen Eight-Four*. The idea is simple: stuff a bunch of contestants – house-mates – in a custom-built house and have one evicted every week by public vote. Created by John de Mol, the British version premiered in 2000, after the success of the Dutch version of the show (produced by Endemol). Sixteen years later and the show is still running. Like so many reality television shows, Big Brother spawned endless spin-offs such as *Celebrity Big Brother*…and Mars One.

Big Brother has been very successful, but will a similar recipe work in space? Lansdorp has been asked that question many times in media interviews. His usual response is that the audience will be engaged at every stage of the mission, from the selection to the training, during the trip and finally with the contestants' activities on the surface. But that's a long time – at least ten years – and one thing television producers have learned is that audiences are a fickle lot and have extraordinarily low boredom thresholds. Of course, that point is moot if Lansdorp can't get the funding and solve the myriad technological challenges. As we'll see later, the chances of that happening are slim to none. And even *if* the technological and financial challenges could somehow be overcome, what about the contestants, er… crew? Lansdorp argues that applicants will be selected carefully, but as we shall see, nothing could be further from the truth; a cursory glance at the list of the final 100 contestants would give even the most charitable astronaut selector sleepless nights. And radiation? Lansdorp's solution is a water tank. Read on to see why that won't work. As for making the trip himself, Lansdorp, who admits his girlfriend thinks he's crazy, hopes to go one day, but acknowledges he isn't cut out for the trip.

Given the vague concept, it wasn't surprising that many people cried 'hoax' and 'scam' when Mars One was announced. It was a reasonable reaction. After all, how on Earth could a group of fantasists, led by someone who made his money selling his wind-harnessing energy company, pull off a trip to Mars for the budget price of $6 billion with no viable business plan? Cue online discussion forums, blogs and experts weighing in; among them Nobel prize-winning Dutch physicist, Dr Gerard 'T Hooft, who said:

"This is an extraordinary project with vision and imagination. My first reaction was like anyone – 'this will never happen'. But now look and listen more closely; this is really something that can be achieved."

But not all experts were convinced. Dr Chris Welch, director of Masters Programs at the International Space University (ISU), wasn't so upbeat in his assessment:

"Even ignoring the potential mismatch between the project income and its costs and questions about its longer-term viability, the Mars One proposal does not demonstrate a sufficiently deep understanding of the problems to give real confidence that the project would be able to meet its very ambitious schedule."

Chris is right. At the time of the announcement, Mars One had a vague idea of the suppliers but had not secured sponsors (four years later and the story hasn't changed).

And even if they had, how could a bunch of amateurs pull off a mission in the mid 2020s that NASA has penciled in for the late 2030s? Part of the answer to that is that Mars One isn't bringing any of its contestants back.

Figure 1.3   Inspiration Mars: a budget trip to the Red Planet but without the landing. Credit: Inspiration Mars

MISSION DESIGN

There have been all sorts of manned Mars mission architectures bandied about over the years, one of the most recent ones being Inspiration Mars (Figure 1.3), a venture that plans to send a couple of astronauts on a 500-day Mars fly-by. No landing, but at least the crew gets to come home. Hopefully. But that's not the case for the Mars One astronauts. Here's how *their* mission is planned:

First, sometime in the early 2020s, a demonstration mission will make its way to the Red Planet to prove some of the, as yet unspecified, technologies that will be needed for the human mission. At about the same time, a communications satellite will be launched that will be placed in a stationary orbit around Mars. This satellite will provide almost round-the-clock communications between Mars and Earth. Joining the demonstration mission and the communications satellite will be a rover that will provide a mode of transport for the landers. The rover's first job will be to drive around to scout out a prime location for landing the crew habitats. Once that job is done, the rover will go about its second job of preparing the surface for the arrival of the first cargo missions and also clearing areas for the solar panels. Sometime after the rover has landed, a second communications satellite will be launched into orbit around the Sun. This satellite, which will be placed at the L5 Lagrangian point, will ensure 24/7 communication between the two planets, together with the first satellite.

Figure 1.4   Concept Mars One habitat. Credit: Mars One

The next step will be the launch of six cargo missions, carrying a second rover, two habitats, two life support systems and a supply module. This cargo will land about 10 kilometers away from the outpost (Figure 1.4). The first job will be performed by the rover, which will collect the first life support system, position it in the right location and deploy the solar panels for the system. Once that is done, the rover will collect the other cargo units and deploy more solar panels. Then the rover will go about the task of connecting the habitats and the life support system, before finally activating the life support system. After activation, the rover will deposit Martian soil into the life support system, so that water can be extracted by evaporating subsurface ice particles. The evaporated water will be condensed into liquid; some of it will be stored and some used to generate oxygen. Looks great on a PowerPoint slide!

With the outpost ready and waiting, it will be time for the crew to start their trip. By this time, the life support system will have produced a breathable atmosphere, and generated about 3000 liters of water and about 250 kilograms of oxygen. The habitat will also be radiation-proofed thanks to the work of the very busy rover, which will have covered the habitats with Martian soil. With the green light on the life support systems and on the habitat, the first step of the manned phase will get underway with the launch of the Mars Transit Vehicle (MTV). First, a transit habitat, a Mars lander and an assembly crew will be launched into Low Earth Orbit (LEO). Here, the assembly crew will dock the Mars lander with the transit habitat, before kicking back and waiting for the next elements – the propellant stages. About three to four weeks later, the propellant stages will be sent into LEO, where they will be connected to the lander-habitat configuration. With the spacecraft fully assembled, it will be time for the crew to launch. Once in orbit, the Mars One crew will exchange places with the assembly crew and prepare for their departure. Once all the systems are checked out, the rockets will be fired and the whole kit and caboodle launched on a Mars transit trajectory. Shortly after the departure of the Mars One crew, a second cargo mission will be launched en route to Mars.

About 24 hours before landing, the Mars One crew will move from the transit habitat to the landing module. The landing module will detach from the transit habitat and, by some miracle, they will land on the surface of Mars. After spending a couple of days adjusting to Martian gravity, the crew will be collected by the rover and transported to the outpost, where they will spend a few more days recovering and settling into their new home away from home. Then they will go about the jobs of deploying more solar panels, installing connecting units between the habitats and establishing food production units. A few weeks later, the cargo for the second crew will land and the first crew will collect the cargo and install whichever elements need to be installed in preparation for the arrival of the second crew. A couple of years later, the second crew will depart Earth, followed closely by cargo for the third crew. This sequence will be repeated several times until gradually a settlement will be established.

THERE IS NO PLAN!

It's a ridiculously ambitious plan that requires all sorts of technologies that have yet to be developed, never mind built or even field-tested. Initially, the plan was for Lockheed Martin to build the rover, while Surrey Satellite Technology would build the communications satellite. Both companies have confirmed that contracts were initiated, but since 2013 there has been little or no progress on these elements, *because there has been no money!* The same goes for the life support system, the study for which was supposed to have been conducted by Paragon Space Development Corporation. Paragon was also the company penciled in to design the Mars suit. The launchers? If you look closely at the Mars One concept art, the habitats bear an uncanny resemblance to the Dragon. This isn't surprising, because the notional Mars One lander *is* a Dragon (Figure 1.5) and the notional launcher is the Falcon Heavy, although SpaceX has no contract with Mars One. And the transit module? One potential supplier mentioned was Thales Alenia Space.

As we know, Mars One has received lots and *lots* of criticism relating to medical, technical and financial feasibility, and we'll get to those later in this book. For now, let's focus on the viability of the mission plan. Er… what mission plan? The reality is *that there is no plan*. Just chatter and misguided statements by delusional bloggers. You would think after four years Mars One would have some vague idea about how to ferry their contestants to Mars, but zip! Nothing. The company wrote a whole book about the venture and somehow managed to sidestep mission architecture, possible launchers, habitats… in fact just about *everything* needed to go to Mars. Instead, they included chapters about the politics of settling on the Red Planet and warbled on about the impact on the participants of being filmed during the selection and training process. Please! The bottom line is that almost all the technology that forms the Mars One mission architecture – such as it is – does not exist. Furthermore, the budget is way too low, the timeline is off by several decades, there is no media corporation that has bought exclusive media rights, the business plan is sketchy at

Figure 1.5   Using retro-propulsion to land on Mars. SpaceX is getting along with the program of perfecting this technique, which is essential for landing safely on the Red Planet. Credit: SpaceX

best, and there are way too many unanswered questions on the subject of keeping people alive en route, never mind on the surface. In short, all the optimism and Nobel Prize winners in the world won't get this puppy off the ground. Sure, Mars One has had plenty of people excited, but the odds are stacked against it. You can talk about the pioneering spirit of the human race and prattle on about how such a mission inspires people until the cows come home, but the harsh and undeniable reality is that a manned mission to Mars – one-way or return – remains a distant dream for those on a Walmart budget. Sure, an underdog non-profit organization with no track record of space exploration and a wildly ambitious plan to land humans on Mars will get tongues wagging, but it doesn't alter the immutable fact that death by radiation sickness is an unpleasant way to die. How successful will a reality TV show be after the Mars One contestants have vomited themselves to death inside the cramped confines of a spaceship?

> "There are definitely dangers associated in going to Mars, but with the right training I am totally willing to face them. I have to imagine that Mars One is smart enough to take that into account while choosing their astronauts."

> *Erica Meszaros*

"Obviously this is something that has captured the public's imagination, and Mars One obviously has a great PR team, but space engineering obeys the laws of physics, not PR."

Anu Ojha OBE,
Director of the UK National Space Academy Programme

Imagine all you want Erica, but you might want to listen to Anu! Now there may be some readers who will argue that if we don't try we won't succeed, but without a viable plan, without financial support, without rugged and tested technology and without a sure-fire way of protecting its contestants against killer radiation, Mars One won't even get to try. Sorry to rain on your parade, but the following chapters explain why.

2

Is this ethical?

Figure 2.0   Shackleton's legendary ship, *The Endurance*. Public domain

© Springer International Publishing Switzerland 2017
E. Seedhouse, *Mars One*, Springer Praxis Books, DOI 10.1007/978-3-319-44497-0_2

Before we get started on the ethical aspect of Mars One, it should be pointed out that the idea does not exclude the possibility of some contestants returning. Unlikely, but possible. Also, Mars One offers the following words of assurance on its website:

"All those emigrating will do so because they choose to. They will receive extensive preparatory training so that they fully know what to expect. Astronauts that have passed the selection process can always choose not to join the mission at any time and at any point during preparations. Back-up teams will be ready to replace any crew member that drops out, even at the very last minute. Our first and foremost priority is to offer the people on Mars as high a quality of life as we can, which encompasses the following:

- Unlimited access to email and other communication channels to keep in touch with friends and family back on Earth
- As many exploration and experimentation opportunities as are available
- The means to build and develop as much as they can themselves. They can work on the expansion of their Mars base and use the new rooms as they wish."

Comforting words, but in the 21st century, a trip to oblivion strikes many as rather extreme, even if it is painted as a bold and unique venture. Those on the Mars One side of the fence argue that such trips have been commonplace throughout human history, citing the example of the thousands of Europeans who moved to Australia with everything they owned. No return ticket for these guys.

Figure 2.1   A short haired guinea pig. Credit: Jg4817

CONSENTING TO DIE

What about the ethics of using the Mars One contestants (Figure 2.1) as guinea pigs? This question is raised as a result of an answer Lansdorp gave in a Q&A with *The Guardian* newspaper. When asked about what research the colonists will perform on Red Planet, the Mars One CEO replied that they would provide an "interesting research topic in itself for physiologists." Really? So the contestants will be research subjects? Nuremberg Trials anyone? Since the Nuremberg Trials, it has been a requirement for scientists who plan to use humans as research subjects to follow certain ethical rules. These rules state that research proposals must be submitted to ethics committees for approval and that *all risks* must be identified. Once the research has been approved, scientists must obtain informed consent from each research subject. Will Mars One meet these conditions? Well, as far as we know, the Mars One plan to offer its contestants as research subjects hasn't been subject to ethical scrutiny, but the question of ethics goes beyond informed consent. Consider the fact that such a mission subjects the contestants to an environment in which reduced bone density, muscle atrophy, radiation sickness and death are guaranteed outcomes. Guaranteed. Consider also that these maladies will occur millions of miles away from a critical care facility and that Mars One has no plans to treat these conditions. It has no plans full stop, but we'll get to that later in the book.

Before we consider the ethical implications of showing contestants dying on screen, let's stay with the research issue for a moment. What about the psychological problems and the human dynamic during such a mission? As we will see in Chapter 4, the contestants will not have the breadth of experience that today's astronauts can draw upon. Consequently, they will not have faced the arduous training that enables those astronauts orbiting the Earth on the International Space Station (ISS) to deal with the problems of social isolation and confinement. And without that training, combined with the duration of the Mars One mission, it is inevitable that mental health issues will arise. While this will be gold for the reality TV viewers, for those stuck in a spaceship having to deal with the drama 24/7, the consequences won't be pretty. Ethically, if a contestant was to have a severe mental health issue, the reasonable course of action would be to suspend television coverage and get the psychological support team involved. But the TV executives would almost certainly deep six this idea and who could blame them? After all, their company may have paid hundreds of millions of dollars for television rights, and reality TV demands spectacle and drama: the more dramatic the better. No, you can be sure as eggs are eggs that any such drama will be milked for all its worth. Counseling sessions? Are you kidding? Sure it will be an invasion of privacy, but this is reality TV millions of miles away! And making the spectacle that much more dramatic will be the communications delay. Imagine if one of the crew has a meltdown (see sidebar) and an altercation ensues. The audience watching at home will be viewing the fisticuffs as it happened as long as 20 minutes earlier. Who's to say one or more of crew hasn't been killed in the meantime? Great viewing potential! Responsibility you say? Take a look at any one of the myriad reality TV shows out there and ask yourself if the notion of responsibility has ever been applied? Ever. Exactly.

Figure 2.2   Lisa Nowak. Credit: NASA

Even astronauts have meltdowns

It took NASA astronaut, Lisa Nowak (Figure 2.2) 12 days, 18 hours, 37 minutes and 54 seconds to secure her place in one of the world's most elite clubs, when she flew aboard the Space Shuttle Discovery during mission STS-121 in July 2006. It took her about 14 hours to destroy it. That was how long it took the 43-year old mission specialist to drive the 1500 kilometers from Houston, Texas, to Orlando, Florida, carrying with her a carbon-dioxide-powered pellet gun, a folding knife, pepper spray, a steel mallet and $600 in cash. Nowak had discovered that Colleen Shipman, a US air force captain, was flying in from Houston to Orlando that night and Nowak wanted to be there to 'scare her' into talking about her relationship with the man at the center of a love triangle. That man was Bill Oefelein, who underwent astronaut training with Nowak, and like her, went into space for the first time in 2006, although they had never flown together.

Shipman allegedly saw Nowak, whom she had never met before, wearing a trench-coat, dark glasses and a wig, following her on a bus from an airport lounge to a car park. Afraid, she hurried to her car. She could hear running footsteps behind her and as she slammed the door Nowak slapped the window and tried to pull the door open. "Can you help me, please? My boyfriend was supposed to pick me up and he is not here," Nowak was alleged to have pleaded. When Shipman said she couldn't help, the astronaut started to cry. Shipman wound down her window, at which point Nowak discharged the pepper spray. Shipman drove off, her eyes burning, and raised the alarm. Nowak was subsequently charged with attempted first-degree murder in what quickly became the most bizarre incident involving any of NASA's active-duty astronauts.

To say the group to which Nowak belonged (her assignment to the space agency was terminated by NASA on March 8, 2007) is select is an understatement. Up to 2007, NASA had selected just 321 astronauts since the US agency began preparing to go into space in 1959. She had been subjected to NASA's rigorous screening process and had trained for 10 years to cope with the intense stress of spaceflight before her mission. Like all the other astronauts, Nowak had been subject to extensive psychiatric and psychological screening, all of which made her behavior incomprehensible.

To many, the Nowak scandal called to mind every bad science fiction movie where they send unstable characters into space. Others argued that NASA should have noticed the signs of Nowak's unraveling. These people might have had a point, but you have to remember that people in highly stressful jobs are generally over-achievers, who put a high value on performance and a low value on self-care beyond that required to perform the job. These types – astronauts in this case – do a great job ignoring and denying signs of fatigue, either physical or psychological, just like polar explorers. Instead, they assume a machine-like thought process to deal with any problems. But the human brain isn't just a thinking machine, it is also the seat of emotions, and the suppression of emotions plays out in the battlefield of the subconscious mind. That suppression and the associated physical and psychological damage eventually surfaces in skewed thought processes and actions, which

(continued)

(continued)

is exactly what happened to Nowak. From our perspective, Nowak's actions appeared crazy, but her perception of her actions appeared to be a logical way to resolve her problem.

Captain Nowak's drama played out in an airport parking lot. Imagine a comparable scene on a spaceship en route to Mars carrying a group of under-trained reality show contestants.

THIRD QUARTER EFFECT

On the subject of psychological breakdown, let us consider a condition known as The Third Quarter Effect. This condition has its origins in the behavior observed in personnel who have spent time in Isolated, Confined and Extreme Environments, or ICE. ICE studies have repeatedly shown that when you put people in extreme environments for long periods of time, you can expect to see decrements in performance, which can in turn lead to less than favorable mission outcomes. For example, we know that on at least three occasions Russian missions have had to be aborted because of psychological problems suffered by the crew; possibly as a result of the phenomenon known as The Third Quarter Effect. The term was coined by Albert Harrison who, in his book *Spacefaring; The Human Dimension*, suggests that long duration space and submarine missions are characterized by three stages. In the first stage, crewmembers more often than not experience excitement and anxiety, while in the second stage the most common symptoms are boredom and depression. In the third stage, many crewmembers become increasingly aggressive and emotional. This is the phase that Harrison dubs the *third quarter phenomenon*, because it usually occurs just after halfway through the mission.

The interesting feature of the Third Quarter Effect is that the three stages occur no matter what the length of mission, whether this be four weeks, four months or four years. As you can imagine, at the beginning of any expedition the crew will be enthused about what lies ahead, but this feeling of excitement inevitably gives way to boredom as the routine sets in (Figure 2.3). For the first half of the expedition, the crew tries not to think too often about how much more time is left before they get to go home, because the finish line is just too far away. But as the mission grinds past the halfway point, the end of the mission becomes tangible. And when a little more time has passed, the crew inevitably begin to look more closely at the calendar and start thinking about returning home to their families and getting back to a normal life. But at the same time, another thought arises, and that is the fear that perhaps the mission will be extended, or maybe something terrible will happen and there will be no return. Not surprisingly, these thoughts may be manifested as the fear and tension that are hallmarks of the Third Quarter Effect. I've seen this happen in the military when I was deployed to the jungle in Belize for six months between 1987 and 1988. To begin with, everyone who hadn't been to the jungle before was excited, and this excitement was sustained for quite some time as we conducted jungle patrols and spent the weekends diving on the islands. But as the halfway mark came into view, the mood gradually changed. With two months left on the clock, the increasing tension came to the boil

Figure 2.3 Shackleton and his crew. Public domain

one evening when some of my platoon went on a bender in San Ignacio, a sleepy town straddling the border between Belize and Guatemala. Soldiers got drunk, bottles were thrown, bones were broken, skulls were split open, one soldier nearly drowned and the military police was summoned from Belize City.

While mission meltdowns haven't been quite as dramatic in Low Earth Orbit (LEO), there have been a number of instances of the Third Quarter Effect in space. In some cases, astronauts have vented their mission frustration and tensions by scapegoating, with mission control being an easy target. Skylab 4 comes to mind. Actually, the strategy of dissipating tension on an outside target is a lot safer than going on a bender and picking on a crewmember, but it is still disruptive. And the longer the mission, the more pronounced the Third Quarter Effect is. Now imagine a very long mission where the supportive role of mission control is reduced by communications lag. Worse still, imagine a mission that *has* no third quarter. Or no quarters at all! Step forward Mars One. During such a mission, our contestants will be without any of the psychological buffers that every crew has had since Gagarin. No real time interaction with family. No instant access to mission control. No option of returning home. Take any survival situation over the past 100 years and you will see one common thread about the will to survive (Figure 2.4) and it is the prospect of a reunion with friends and family. It is this that galvanizes the behavior necessary to keep going despite insurmountable odds. Remove the element of social connectedness and you have a sure-fire recipe for a less than optimum outcome. Yes, yes, yes, the one-way mission makes sense for any number of technical, engineering and financial reasons, but the psychological handicap could be so great that it may put the whole mission at risk. Ethical? Hardly.

Figure 2.4   The Mawson expedition. Public domain

Now, the Mars One crowd will try to argue that the one-way trip isn't really that extreme and besides, thousands of Europeans embarked on a similar one-way trip in the 1800s, when they left their home country without a return ticket, bound for Australia. Such an argument is flawed logically, because the Mars One astronauts will be emigrating for very different reasons than their historical counterparts. You see, back in the 1800s, immigration was driven by Britain's requirement for a penal colony, which was populated mostly by convicts and by settlers hoping to make money from a gold rush. Mars One, on the other hand, is asking for volunteers to take part in a venture for historic reasons rather than searching for a better standard of living, so it is wrong to use Australia's immigration history as a parallel to a one-way trip to Mars.

So how does Mars One get away with the one-way branding of their mission? Numbers. Pure and simple. In many ways, Mars One uses the sheer number of applicants as justification that the plan is a viable one, despite mountains of evidence to the contrary. But this is exploitation of the enthusiasm of the contestants, who are led to believe that if they are selected they will be part of history by establishing the first human colony on the Red Planet (they won't be first because SpaceX will likely get boots on the surface before them). Now there are many who would say good riddance to anyone gullible enough to think that Mars One is anything but a flamboyant attempt to win a Darwin Award, but the fact remains that every space mission to date has offered a way back to Earth.

DYING TO DIE ON REALITY TV

Another contentious ethical issue is turning the one-way venture into a reality television show. Not only must contestants give up any chance of returning to Earth, but they must also face the ordeal of having millions of people watch them en route. Nothing wrong with another coma-inducing reality television show, you may think. After all, there are plenty of people happy to kill their brain cells by watching *Keeping up with the Kardashians*, so why not kill a few more by watching contestants fly to Mars? Well, the answer lies in the very limited field experience we have of humans living for long periods in space. You may have heard about the one-year mission that took place on board the ISS between 2015 and 2016, when Scott Kelly (Figure 2.5) and Mikhail Kornienko spent 12 months on the orbiting outpost. That's a fair stretch, though Kelly and Kornienko's mission fell short of the record held by Valery Polyakov, who spent 438 days on board the Russian space station *Mir*. But 438 days is much, *much* shorter than the six-month journey and the lifetime that Mars One contestants will be spending on the Red Planet. Granted, Mars provides a gravitational field that is 38 percent the strength of Earth's, so the Mars One group will be afforded partial protection against some of the deleterious effects of reduced gravity, but we have no idea how long duration exposure to deep space affects the human body, so the Mars One contestants will effectively be lab rats in a show that will inevitably take a nosedive (see Chapter 5).

Now let's consider the ethical concerns about how this Red Planet boondoggle is supposed to be funded. Hypothetically, Mars One reckons it will cost about six billion dollars to establish a Mars colony and sustain the lives of the contestants. Now, anyone who has worked in the space industry will tell you that this figure is spectacularly unrealistic. Almost as unrealistic as expecting that the Mars One reality television show will generate about four billion dollars in revenue, but let's indulge the Mars One organization and accept their wild assumption that millions of people around the world will watch the show. But, as anyone who watches reality television knows, these shows often have short life spans, so what happens when people get bored with watching crewmembers pressing buttons, flicking switches, reading checklists and dealing with radiation sickness? Well, when the money dries up, the chances are the supply chain will dry up too. No more food, oxygen or medical supplies. Mars One will have to explain to their contestants that the mission will have to be abandoned due to lack of funding. What will the contestants do? This is where it could get interesting. Don't forget that the contestants will have control of the cameras. Perhaps they could switch them off? Perhaps they could film the first death on the Red Planet and temporarily resurrect interest in the enterprise, although it probably wouldn't be enough to save them.

On the subject of saving a failing reality television show, let's talk about sex. There is always the remote possibility that before the end of the first season, one of the crew could have been 'knocked up', in which case there would be a ratings bonanza. In reality, this is extremely unlikely, because the crew will have long since been sterilized by all the radiation. But let's indulge. Is it really ethical to give birth on the surface of Mars? Probably not. We have absolutely no knowledge of how a human fetus will develop in a reduced gravity environment. What will Mars One do to prevent such an event? Will they sterilize

Figure 2.5   Scott Kelly. Credit: NASA

the crew? Will they provide contraceptives? (Figure 2.6) Will these contraceptives have been tested in space? By whom? And what if the surface of Mars turns out to be an unfavorable one to conceive? Well, that torpedoes the idea of establishing a colony doesn't it? In fact, the Mars One base will perpetually face extinction unless there is a steady stream of humans from Earth. Hmm. Seems someone hasn't thought this through, although Mars One at least acknowledges the challenges of interplanetary sex:

> "The Mars settlement is not a suitable place for children. The human ability to conceive in reduced gravity is not known, neither is there enough research on whether a fetus can grow normally under these circumstances."
>
> *Mars One website statement on the issue of sex and the*
> *ethical consideration of conceiving children on the Red Planet.*

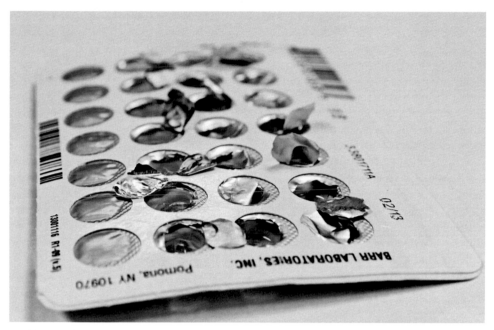

Figure 2.6 Contraceptives: These probably won't be needed, but you never know. Best be prepared. Credit: Bryan Calabro

Another aspect that seems to have been given short shrift is the issue of social isolation. The Mars One contestants will be the most isolated humans in history. And because of the distance from home, real time interaction will be impossible. In fact, for the rest of their lives, the only real time interaction the Mars One contestants will have will be with each other (and with any subsequent arrivals if, by some act of divine intervention, they manage to stay alive long enough to meet them). We know from decades and *decades* of research, and by combing through the annals of Antarctic exploration, that prolonged social isolation may lead to mental illness, often manifested by depression, anxiety, chronic fatigue, insomnia and emotional instability. It's a problem that the Mars One team has considered and here is what one of their experts thinks:

> "It all starts with attitude. Think of it. When a person finds herself, or himself, on Mars, with no way of being able to come home, and potentially questioning the decision that they have made, what is going to ground them in the choice they have made?"

Attitude! Of course! Why didn't I think of that? *Attitude* is the solution to all those problems of homesickness and dysphoria. Never mind that even the most highly trained astronauts have suffered feelings of isolation. Never mind that even some of the boldest and most fearless explorers have succumbed to homesickness. Apparently, it was all a question of *attitude*. Thank you Mars One! But just suppose one of the Mars One contestants does suffer a major breakdown. What then? Who will be responsible? Perhaps applying the right attitude can also help the Mars One contestants deal with life indoors? Because from the moment they land, these would-be colonists will be stuck inside a bunker, unable to venture out on account of the lethal radiation, the unbreathable air and the paralyzing cold. And this confinement will take place in habitats that provide about 50 square meters per person. Now imagine a regular day in your life. Imagine all the sensory experiences and all the different environments you take for granted. The Mars One contestants will live out the rest of their – albeit short – lives with just a very, very small fraction of this. As you might expect, extended periods of confinement provoke the same problems as social isolation (Figure 2.7): anxiety, boredom, depression, and cognitive impairment – take your pick. And then there's the loss of privacy. Don't forget that the movements of the Mars One contestants will be observed 24/7. How do you think being under constant surveillance will affect this group? Don't forget that they will already be under tremendous stress and we know from countless research studies that, even under the best circumstances, being watched can itself cause fatigue, anxiety and stress. The Mars One team have yet to make any comments about the effects of social isolation and confinement. Perhaps they're secretly hoping that these factors will conspire to bring about meltdowns in the crew, thereby generating tension and a boost in ratings? Who knows?

Figure 2.7   The Cupola of the International Space Station. Credit: NASA

"The only type of person that I can honestly picture myself having trouble living with is someone who is consistently negative, or worse, not a team player. Personally, I'm a glass-half-full girl. I have an endless supply of optimism and I approach challenges and obstacles with humor. I never call something a problem. Instead I put my nose to the grindstone and figure out how I can best contribute to a solution."

Kellie Gerardi, Mars One Final 100 Candidate

"I don't think there is a specific personality that I would not be able to get along with at all. I think every person has a diverse mixture of personality traits, so there is always something that I can find to relate to with a person. If there is a will there is a way. I am highly adaptable and my humorous nature does help me relate to others. I think if a person just takes time to get to know another person, there is always a way to coexist. Most people do not seem to know their own limits and who they are, but if they are open enough to others, they can also discover themselves. Traits that might not agree with my personality are a lack of curiosity, no sense of adventure, no ability to think and act in stressful and extreme situations, an unwillingness to cooperate, and dishonesty."

Andreea Radelescu. Mars One Final 100 Candidate

REAL EXPLORERS

What does Mars One say in its defense? Well, one of the standard responses has been to compare the conditions faced by contestants on the Red Planet with the conditions faced by explorers in the 'Heroic Era of Antarctic Exploration'. The argument is that humans have suffered appalling hardship before, so presumably they can do it again. This argument is weak in the extreme. First of all, explorers such as Amundsen and Shackleton knew they had a reasonable chance of making it back alive, rather than the prospect of living in a confined bunker and an early death. Second, these great explorers of old *did* suffer appalling hardship, but those conditions would probably be judged unacceptable in the modern era. And third, even if these conditions were deemed acceptable, chances are the explorers of today wouldn't be up to the task. You see we live in a mollycoddled society in which the comfort blanket gets comfier every day. There are still explorers, but the dangers faced by this group today are incomparable with those faced by Mawson and his ilk (Figure 2.8). Explorers today are rarely more than a satellite phone call away from rescue. In short, today's explorers, however bold they may appear, would still be woefully unprepared for the harsh realities that would await them on the Red Planet. And this unpreparedness will be costly in the extreme. Ask yourself if you would allow your children to watch a reality television show in which the contestants are dying from radiation sickness, bleeding from every orifice, and/or suffering from profound confinement-induced paranoia. This isn't so much about being risk averse or cowardly; it is about being smart. And being smart does not mean sending a woefully unprepared crew on a suicide mission,

Figure 2.8   Douglas Mawson. Public domain

because one thing we've learned over the years is that nothing weakens political will quicker than a disaster.

> "We wanted to get some idea of what will happen when a motivated, high-performing crew is confined in a spacecraft-like environment for a full 17 months, simulating a mission to Mars and back. The assumption has been with the six-month space station missions that anybody can tolerate them. OK, you have trouble with your sleep or something but you're only up there six months, it won't last. If you go on an exploration mission, you'll adapt. But our study shows that's not true. The people did not adapt. In fact, these problems just cumulatively created a greater and greater physiological and behavioral burden on the crewmember."
>
> *David Dinges, University of Pennsylvania's Perelman School of Medicine*

But let's get back to the woeful unpreparedness of today's current crop of explorers. As you can see from David Dinges' quote, there has been plenty of research conducted on the subject of confinement and social isolation, and the results have been less than encouraging. That quote refers to the Russian-European Mars500 boondoggle, in which three Russians and three Europeans were sealed inside a tin can for 520 days to simulate a mission to the Red Planet (Figure 2.9). During the ground simulation, the crew suffered from sleep disorders and crew communication issues – neither of which bode well for longer journeys – although such problems were judged to be manageable since the crew knew there was an endpoint.

Figure 2.9 The Mars500 guinea pigs. Credit: ESA

Any normal mission comprises an outbound, a landing and an inbound component. Any psychologist will tell you that if you have to remove one of those components, it should be the second, which is exactly the thinking behind Dennis Tito's Inspiration Mars idea (Figure 2.10) – a fly-by of the Red Planet which will take about the same length of time as Mars500. Not only is Inspiration Mars a whole lot more feasible, cheaper and ethically sound than Mars One, the mission would also test the technology required for the outbound and inbound phases of the mission.

Figure 2.10 Inspiration Mars banner. Credit: Inspiration Mars

RULES

The business of space exploration is a risky one, in which rockets explode with alarming regularity (SpaceX in 2015, Orbital, Soyuz, etc., etc.). But, as we've discussed in this chapter, an exploding rocket may not be the worst thing that could happen to the Mars One contestants. They could suffer agonizing deaths brought on by radiation sickness, or they could all simply go mad from extreme confinement and isolation. Or both. Consider the following scenarios:

Scenario 1
Imagine you are one of the (un)lucky four to have been selected as the first Mars One crew. You are three months into the mission when the spacecraft suffers a rapid decompression caused by a micrometeorite puncturing the vehicle's skin. Thanks to immediate action by the crew the hole is repaired, but life support consumables have taken a hit. In fact, the oxygen tanks now only have enough oxygen to support three crewmembers for the remaining three months that it will take for the spacecraft to reach Mars. According to the very best assessment made by mission planners, immediate action must be taken to prevent Loss Of Mission (LOM) and/or Loss Of Crew (LOC). In short, that action means one of the crew must be euthanized to ensure that at least three make it to Mars intact. What does the commander do? Should he/she be sacrificed? After all, perhaps he/she is the largest member of the crew and therefore consumes the most oxygen. But if the commander is killed, who is going to command the mission? Should Mission Specialist #1 (MS1) be sacrificed? But MS1 is the crew doctor, so who will treat the contestants after landing? Perhaps the engineer should be terminated, but then who will repair all the systems for the remainder of the trip? And since it was the quick actions of the engineer which saved the mission following the rapid decompression, it would be unfair to kill him wouldn't it? The scientist then? But if the scientist is killed, who will make the discoveries on the surface? It's a tough call and it has to be made immediately, otherwise Mars One is looking at a definite LOM. But which guidelines are followed and who makes the decision? Is it the commander? Mission control? Bas Lansdorp? Perhaps the crew should just draw straws and be done with it?

Scenario 2
Let's take another scenario. This time, the crew are two days away from the critical Entry, Descent and Landing (EDL) task when mission control receives news that the pilot's wife and two kids have been killed in a car accident, while driving to a television studio to be interviewed for the Mars One reality television program. This is particularly bad news because the pilot's back-up died from radiation sickness just three days previously. Does mission control tell the crew and the pilot immediately or do they wait until after landing? And what happens if that information is leaked and the pilot receives the information before the EDL? Does that mean the mission is aborted?

Scenario 3

Three months after landing, the crew has been reduced to just two crewmembers, after one contestant died from radiation sickness and the other committed suicide by blowing herself out of the airlock. Worse, the consumables have taken a hit after food was spoiled following a fuel leak and oxygen was inadvertently vented following a life support system malfunction. There is now only enough food for one crewmember to survive until the next resupply mission and even that will be tight. Once again we're faced with a euthanasia dilemma, but with a twist: in this case it isn't just one crewmember having to kill another, but what to do with the corpse after the kill. Remember, food supplies are below critical, so it would make sense to eat the corpse. Are there any procedures for this? What would you do Mars One? It's your call.

THE RAGGED EDGE

A manned mission to Mars lies beyond the ragged edge of what is achievable today, and it will do for many, *many* years to come, unless Elon Musk has something up his sleeve that he's not telling us. And when astronauts do finally venture into the void, they will embark on a mission in which many, *many* things will go wrong. Best be prepared then when it comes to the issue of ethics, so crewmembers at least have a framework to guide them in making lifeboat decisions. It will also help if crews know the answers in advance to the types of scenarios depicted previously, because trying to make decisions when everyone is panicked is only going to make an already bad situation worse. And it's not as if advance crisis planning hasn't been done before. Remember the Apollo missions? In the event that Neil Armstrong and Buzz Aldrin never made it back to lunar orbit, President Nixon had the following speech ready:

> "Fate has ordained that the men who went to the Moon to explore in peace will stay on the Moon to rest in peace. For every human being who looks up at the Moon in the nights to come will know that there is some corner of another world that is forever mankind."

Now you may think that I'm exaggerating here and we needn't worry about nasty events such as rapid decompression and radiation sickness. After all, we've been sending humans into space for decades now and we've been fairly successful at it. Correct, but a manned mission to a space station or the Moon does not even begin to come close to the reality of the myriad problems posed by a multi-year journey to the Red Planet. Which is all the more reason to have an ethical framework in place. Let's consider a few more examples in which such a framework might be needed. We'll begin with one of my favorite subjects – the nonhuman life that lives inside us. Genetically, only about one in ten cells in our bodies are *homo sapiens*, with the rest being comprised of a microbiome. This microbiome is known to have effects on our health, but what we don't know is how long duration space-flight affects it. Is it possible that deep space radiation could cause changes that might manifest as a disease which could ultimately wipe out the Mars One colony? And on the subject of nonhuman life, let's consider alien life. After all, one of the activities of the Mars

One contestants will no doubt be the search for microbes or anything else that could boost the ratings. But what if life is found and the Mars One colonists are deemed to be contaminated? What then? Surely we can't allow the next mission to travel to Mars because the risk of infection would be too great. What ethics protocols are in place for that scenario? Then again, if Mars One decides to go ahead anyway, what authority can stop them? And before we move on from the topic of alien life, what about a baby born on Mars? While the Mars One contestants will have given their consent to take part in the mission, that won't hold true for a baby born on the Red Planet. This baby would be deprived of all the resources available to babies on Earth, so such an event would surely be unethical. But what document says otherwise? The advantage Mars One has when it comes to deciding what is ethical and what is not is that the private space industry is still in its infancy and there are no ethical blueprints for how commercial space crews should behave. But death and serious illness are inescapable in a mission that is as risky as Mars One.

3

Is any of this legal?

Figure 3.0   Credit: NASA

© Springer International Publishing Switzerland 2017
E. Seedhouse, *Mars One*, Springer Praxis Books, DOI 10.1007/978-3-319-44497-0_3

Wherever there are explorers there are lawyers, and there are plenty of legal issues that must be dealt with before humans can go shooting off to Mars. Perhaps one of the most obvious is the legality of actually allowing humans to be selected for a one-way mission. Can Mars One actually get away with this? Well, yes they can and most probably will, and here's why. Most countries have anti-discrimination laws that almost always include a Bona Fide Occupational Qualification (BFOQ) exemption that can be applied to all sorts of jobs, even if that job is being a contestant on a reality television show on their way to a distant planet. Now you may be wondering if any of this legalese applies to Mars One because they are a private entity sending their employees into space; surely this legal stuff only applies on Earth doesn't it? Well, sort of. Sure, Mars One isn't an employer in the traditional sense and an argument could be made that, because of that distinction, the traditional labor laws might not apply. As for legal standards being applicable to astronauts in space, you would be right about some federal laws and regulations. For example, the US Occupational Safety and Health Administration (OSHA) doesn't have any laws that apply to those working in Low Earth Orbit (LEO), or anywhere else in space for that matter. But OSHA does have regulations that NASA employees must abide by while on the ground. And the Outer Space Treaty (OST) and the Moon Agreement? What about these? Well, the OST and the Moon Agreement have plenty to say when it comes to the issues of contamination and ownership, but nothing about regulating the health (Figure 3.1) and well-being of astronauts or reality TV contestants, whether they are employed by NASA or by dubious entities such as Mars One.

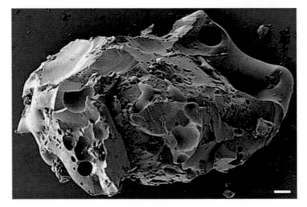

Figure 3.1 Lunar dust caused all sorts of health problems during the Apollo program. Credit: NASA

COLONIZATION

Let's begin by noting that many of the problems the Mars One contestants will face are not *sui generis*. In other words, every manned mission to the Red Planet will face them (Figure 3.2). But what most manned missions probably won't face will be the legal challenges, some of which are examined in this chapter. Before we start, it is important to reiterate the caveats made earlier, namely that the problems discussed are not insurmountable; it's just that Mars One seems not to have anywhere near the resources, capability or

Figure 3.2 Concept Mars mission: There have been dozens of these over the years. Credit: NASA

budget to resolve them. In short, it doesn't have a clue. So what follows is not an attack on the Mars One organization, nor is it an attack on the vision of sending contestants on a one-way mission to the Red Planet. Rather, this chapter scrutinizes some of the more pressing legal issues and how these may affect Bas Lansdorp's deadline and goals.

We'll begin with the issue of colonization. Now, as you know, Mars One is incorporated in The Netherlands and as anyone with even the flimsiest grasp of world history knows, the Dutch don't exactly have a sterling record when it comes to colonization. So one of the first questions we should ask ourselves is whether Mars One can become a colony in the legal sense.

> "A colony is a dependent political community, consisting of a number of citizens of the same country who have emigrated therefrom to people another, and remain subject to the mother country. It is a settlement in a foreign country possessed and cultivated, either wholly or partially, by immigrants and their descendants, who have a political connection with and subordination to the mother country, whence they emigrated."
>
> *Definition from: www.thelawdictionary.org*

Based on the above definition, it would seem that the legal concept of a colony does not apply to Mars One because, as we know, the contestants will be of various nationalities. Not only that, but it could also be argued that since Mars One is a private enterprise, it will not be under the jurisdiction of any government. Seems reasonable, until you start reading through the various articles of the OST; specifically Article VI, which states:

> "States Parties to the Treaty shall bear international responsibility for national activities in outer space, including the moon and other celestial bodies, whether such activities are carried on by governmental agencies or by non-governmental entities, and for assuring that national activities are carried out in conformity with the provisions set forth in the present Treaty. The activities of non-governmental entities in outer space, including the moon and other celestial bodies, shall require authorization and continuing supervision by the appropriate State Party to the Treaty. When activities are carried on in outer space, including the moon and other celestial bodies, by an international organization, responsibility for compliance with this Treaty shall be both by the international organization and by the States Parties to the Treaty participating in such organization."
>
> *Article VI: Treaty on Principles Governing the Activities of States*
> *in the Exploration and Use of Outer Space,*
> *Including the Moon and Other Celestial Bodies*

What might this mean? Well, it means that since Mars One is a non-governmental entity incorporated as a non-profit company in The Netherlands, the company is subject to the jurisdiction of The Netherlands, which means that its contestants will also be considered non-governmental entities under the jurisdiction of The Netherlands. So, if any Mars One contestant ever sets foot on the Red Planet, that person would be establishing a colony on behalf of The Netherlands. Such an event would no doubt trigger a cacophony of controversy about colonialism and The Netherlands might be stigmatized once again. But the ratings! Imagine the ratings! Seriously though, a private company laying claim to an extraterrestrial colony (Figure 3.1) might be construed as a sovereign claim, which might not sit well within the domain of international law.

Figure 3.3 Laying claim to an extraterrestrial colony might be construed as a sovereign claim, which - legally - might rub some people the wrong way. Credit NASA

THE MOON TREATY

Another potential legal headache is the Moon Treaty, also known as The Agreement Governing the Activities of States on the Moon and Other Celestial Bodies. Considered by many to be a failed treaty because so few nations have signed it, this legal document – particularly Article 7[1] which deals with effects on the environment and Article 11 which deals with resource development – could be a problem for Mars One, because The Netherlands is one of the few countries to have ratified it. Of course these issues may become moot if Mars One can't get off the ground, and to do that they need a launch license, which means it will have to perform its activities under American jurisdiction, specifically Title 51, Chapter 509, also known as the Commercial Space Launch Act (CSLA). Now obtaining a launch license is not an easy business, especially when the company applying for the license is an organization like Mars One. You see, if – and it's a big 'if' – the United States issued a launch license to Mars One (Figure 3.4), the license would need to cover many years of launch activities, based on the current Mars One mission plan comprising resupply and manned missions. That's a lot of commitment to be asking for, because any license granted would probably have to be irrevocable since the Mars One colony would be completely dependent on resupply to grow. Just imagine if the

---

[1] Agreement governing the Activities of States on the Moon and Other Celestial Bodies. New York, 5 December 1979; 11 July 1984, in accordance with article 19(3).

launch license was revoked! That would be the kiss of death for all the Mars One contestants, although once again it would push up the ratings. For those of you who follow the goings on of '*Keeping Up with the Kardashians*', you will no doubt have heard all the accusations that many of the episodes are rigged to drive up ratings. Well, it's easy to imagine a similar scenario playing out for the Mars One reality television show. Just kick-start the rumor mill with a story that the launch license has been revoked and watch the ratings climb. Pure ratings gold!

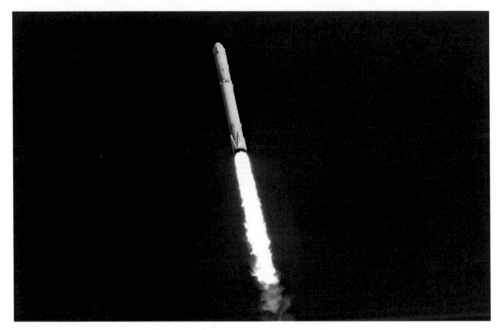

Figure 3.4   SpaceX launcher. Credit: NASA

GETTING A LAUNCH LICENSE

In short, approving a launch license, knowing full well that any revocation would spell certain death to contestants hundreds of millions of miles away, would cause some serious deliberation at the governmental level because of the encumbrance of being responsible for the mission's potential failure. Under these circumstances, the only way the license could be revoked would be if the Mars One outpost failed. If that was to happen, you can be damn sure that would be the end of Mars One, because no government in their right mind would issue a launch license for a space mission that had already killed its participants. And on the subject of killing its participants, it is worth highlighting the issue of risk, because for Mars One to go ahead it will require the approval of the Dutch *and* the American government, and which government is going to green flag a venture in which there is an extraordinarily high chance of failure? If the transit to Mars doesn't kill the crew and if by some miracle they survive the landing, you can be damn sure they won't survive long on the surface. And remember, all of this is being shown 24/7. Imagine the political, public and media kerfuffle, accompanied by millions of Facebook posts and assorted social media outrage, after the public witnesses the agonizing deaths of Mars One contestants. Of course, once again there would be a ratings bonanza, but paralleling this there would no doubt be political repercussions that might ultimately lead to putting the dampeners on *any* future government or private settlement ventures. So, taking all the above into consideration, do you *really* think Mars One will get a launch license? Think again.

So a launch license is a long shot, but life is full of surprises, so let's imagine by some miracle Mars One has the paperwork approved. What other legal hurdles might there be? Well, let's consider the family arrangements. As we'll see in the next chapter, a number of the contestants are married and some have children (see sidebar). What will happen if one or more of these contestants are selected for one of the manned missions? Will they continue to be married, or will the significant other remaining behind file for divorce or have their marriage annulled? Most likely either option will be on the table, because the Mars One contestant will be abandoning their family to start a new life and possibly even hook up with another partner – the aim after all is to start a colony on Mars isn't it? Imagine you're married with a couple of kids and your partner announces they have been selected for a one-way mission to Mars: "And by the way honey-bun, one of the aims of this mission is colonization, so I might be hooking up with someone else. Look after the kids will you luv." What would you do? Most sane people would file for divorce, not only on the grounds that their partner is leaving them but on the grounds they have also taken leave of their senses. But what happens to the kids? Will the Mars One contestant be liable for alimony and/or child support and if so, how will these payments be enforced? The courts tend to take a dim view of those skimping on their alimony and/or child support payments, so a person leaving the planet may be seen as escaping his/her financial obligations, in which case the court could make an order that forbids the contestant from launching. Here's how this conversation might play out:

Court: "Sorry mate, but you're grounded on account that you're in contempt of court. We can't have you clearing off to some other planet and leaving your wife and kids to fend for themselves, can we?"

Mars One Contestant: "But I've been selected for a one-way Mars mission. I'm going to be famous."

Court: "Well that just isn't going to happen. This child support business is serious and I'm telling you that you're grounded."

And so on and so on. Once again, ratings gold. Having said that, one way around this would be to create a trust to pay for the alimony and/or child support. Again, it's just one more question that has to be resolved and another one that Mars One hasn't addressed yet.

A Mars One contestant's take on making a one-way trip

"When we leave for Mars, I will have been married to my beloved wife for thirty-five years. She has been my heart and center for all my adult life. In 2022, my children will be in their late twenties, their lives just beginning to open into their full flower, perhaps ready to have children of their own. I can only imagine how hard leaving them behind will be.

"In the nineteenth century, when millions of people were emigrating from Ireland, although they might receive a rare letter from the Old World, they also faced the near-certainty of never seeing their friends and relatives again in the flesh. Around this time, they began to celebrate what came to be known as an 'American Wake'. They would have a final party in which those leaving and those staying behind could share a final drink, have one last dance together, and say a final farewell. I think my loved ones and I will need the closure that only a 'Martian Wake' will provide. After that, email and laggy video calls with my family and friends will have to suffice to keep love strong.

"While it is a difficult thing to be sure of (and also a hard thing to admit), I think there are certain emotions I feel less keenly than most people, particularly anger and sadness. And although, previously, I've never had extended occasion or desire to test the theory out, perhaps this quirk of mine will help me deal with the separation from my wife and kids. I hope."

Daniel Carey. Mars One Applicant.

No need to worry Daniel – you'll be around on Earth for quite a while yet. Perhaps your grandkids could make the trip? On board a SpaceX vehicle?

GOVERNANCE

But let's assume Mars One can figure out the divorce problems and let's assume it receives a launch license. The contestants somehow make it to Mars. What then? How will the colony develop? What structured code of governance will be implemented? Which officials will be appointed? How will the safety of colonists be ensured? And how will they ensure the environment is preserved? Let's take the question of legal structure first. Under the Mars One scheme of things, it will be up to the contestants to create a government body which, presumably, will be collaborative and democratic, but then again, who knows? Who will enforce this democratic system is anybody's guess, as is the procedure for prosecution if a contestant steps out of line. Perhaps none of this will matter if astronauts on Mars are not deemed to fall under the jurisdiction of the laws that govern people on Earth. Ask yourself this question: "What are the chances of a small group of average citizens, living in an extremely high risk environment, creating a harmonious society?" Enough said? Perhaps not.

What would the options be for jurisdiction of a permanent habitat on Mars? Well, first of all it is worth highlighting that under the Liability Convention, governments are responsible for their citizens in space. So, one option would be to create a single governmental entity. In the case of Mars One, this would likely be a group of habitats on the surface, populated by contestants who would, in all likelihood, be citizens of as many as four different countries on the first mission. Under the Liability Convention, it would be the government of each national that would be solely responsible for what that national did. The same criterion would apply to the ownership of the habitat, which is similar to the system that works on the International Space Station (ISS). Up there, the orbiting habitat comprises modules belonging to the US, Russia, the European Space Agency and Japan. If any legal incident occurred in any of the modules, it would be the owner government that would retain jurisdiction. Would this system apply to Mars One? Probably not. So what to do? Well, one possible solution is to develop a code of conduct that applies to multinational space habitats. This code could be based on terrestrially accepted principles that apply to civil, criminal and tort law, but they would be applied to extraterrestrial habitats instead. It might just work. Does Mars One have a take on the situation? Well, as a matter of fact, it does. Here is what the oracle on Mars One has to say on the subject (author's italics):

> "Although some of these organizational approaches are less likely to be realized than others, each possibility considered would have profound effects on the discussions that the political and legal apparatus of each international community would bring to any decision-making table. The organizational approach, therefore, has a fundamental impact on the issues of governance and overseeing of activities for creating a settlement on Mars. Furthermore, although the current construct of international space law necessitates that international communities oversee and authorize the activities of all actors under their jurisdictions, *several issues of interest to private enterprise and nongovernmental actors remain to be clarified under the ambit of international space law*. For example, the creation of a Martian settlement includes exploitation of resources, such as valuable new minerals, for their eventual commercialization and possible transport back to Earth, or transforming the very environment of Mars for the creation of a self-sustainable human habitat or for tourism. *The legalities of such actions remain hazy under the current international space law*."

> *Quote taken from the Mars One bible:* Mars One: Humanity's Next Great Adventure: The Politics and Law of Settling Mars: The Need for Change, *by Narayan Prasad. Edited by Norbert Kraft. Benbella, 2016.*

So at least Mars One has some understanding of the lack of legal guidance when it comes to landing a group of contestants on the Red Planet, but the way forward is anything but clear. Now let's move on to the environment. Here's what Mars One has to say on this subject:

> "Mars One will take specific steps to ensure that the Mars environment will not be harmed. The Mars base will be forced to recycle just about everything, and pay close attention to its energy use and minimize the leakage of materials and energy."

COSPAR

According to Mars One, the organization is communicating with the Committee on Space Research (COSPAR) to figure out how best to preserve the Martian environment. One way will be the installation of solar panels and another will be the production and recycling of water and oxygen in the life support system. But what happens if there is an accident? Perhaps one day one of the contestants is a little clumsy and spills fuel on the ground.

Who will be liable for the damage? Who will be accountable? These are difficult questions to answer because there aren't any hard and fast enforceable rules and regulations when it comes to the issue of an independent, internationally-funded venture landing on Mars. Perhaps space law needs to be updated to take into consideration the commercialization of space, because that is essentially what Mars One aims to do by setting up a reality television show on the surface of the Red Planet. How much more commercial can you get? And on the issue of commercialization, what about the not so insignificant matter of ownership? In the OST, space is treated as the 'common heritage of mankind', which effectively denies governments or their citizens from staking a claim to land on any celestial body. But while the matter of land ownership on Mars is a contentious one, a far more divisive issue is the subject of planetary protection. Some provision for planetary protection is stated in the OST, which cites the following guidelines:

> "States Parties to the Treaty shall pursue studies of outer space, including the moon and other celestial bodies, and conduct exploration of them so as to avoid their harmful contamination and also adverse changes in the environment of the Earth resulting from the introduction of extraterrestrial matter and, where necessary, shall adopt appropriate measures for this purpose. If a State Party to the Treaty has reason to believe that an activity or experiment planned by it so its nationals in outer space, including the moon and other celestial bodies, would cause potentially harmful interference with activities of other State Parties to the Treaty shall pursue studies of outer space, including the moon and other celestial bodies, it shall undertake appropriate international consultations before proceeding with any such activity or experiment."

Avoiding contamination by spacecraft sent to other planets is a serious business. Every couple of years, hundreds of scientists get together to conduct COSPAR workshops, tasked with figuring out how the OST provisions can be implemented. But why? Well, it's quite simple really. If a spacecraft accidentally introduced life to Mars, then it would be very difficult to prove life existed before the spacecraft got there and it would also be difficult to figure out if any existing life on Mars hadn't been wiped out by the micro-organisms brought by the spacecraft. So, to prevent such a scenario, the COSPAR scientists make the guidelines stricter and stricter. This means, in an attempt to keep Mars as pristine as possible for as long as possible, spacecraft heading in the direction of the Red Planet must be sterilized to a very, *very* high level (Figure 3.5). To date, we have only had to deal with unmanned probes, but if we start sending humans to Mars… well, preventing contamination becomes nigh on impossible and here's why.

If you think it is possible to go to Mars without introducing terrestrial life, consider first our skin, which has about 1000 species in 19 phyla, making for about a trillion individual organisms. That's just on your skin. On *one* human! Now, stuff four humans into a spacecraft and you have in effect created a highly concentrated package of micro-organisms. And as soon as those humans step outside the spacecraft, contamination is inevitable. Now you may think that our terrestrial micro-organisms wouldn't stand a chance on the radiation-soaked surface of Mars but you'd be wrong, because of a species classified as *extremophiles* (Figure 3.6). These tough little critters can survive in just about any environment.

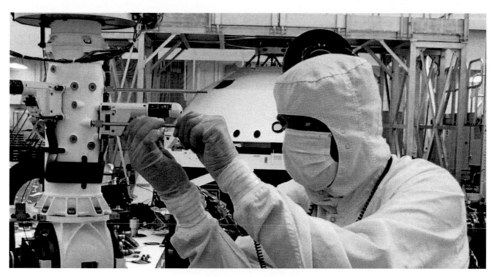

Figure 3.5 Sterilizing spacecraft. Credit: NASA

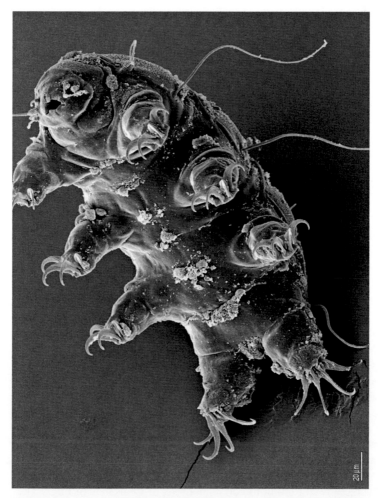

Figure 3.6 An extremophile. Tough little buggers, these. Credit: NASA

A Mars One mission will be bound by COSPAR guidelines, which state quite clearly that the Red Planet must be protected in the event of a crash landing. Given the challenges of Entry, Descent and Landing (EDL) on Mars, a crash landing is a very, *very* strong possibility. If that was to happen, the spacecraft would deposit the remains of four bodies onto the surface. Goodbye pristine environment! And even if the spacecraft landed intact, the chances of not inadvertently contaminating the surface are about as remote as the chance of the spacecraft executing a safe EDL sequence in the first place, but let's talk about that later. The resulting contamination wouldn't just make it difficult for those trying to determine if Mars had life, it would also create a headache for other activities. Take terraforming, for example. Now terraforming is hundreds of years over the horizon, but we may want to do this one day to make the Red Planet more homelike. One way to terraform would be to use photosynthesizing organisms to generate oxygen. But what if Mars has been accidentally seeded with micro-organisms that just happen to *love* oxygen? In fact, they love it so much that they devour any and all oxygen that is created. What happens to your terraforming plans then? After all, Earth needed to go through a lengthy phase when oxygen was being generated but not consumed.

So how does Mars One avoid contamination of the surface? Well, it can't, unless you tweak the mission slightly. Perhaps make it an orbital mission, for instance. No dust storms to contend with and plenty of opportunities for telepresence, although the crew would take a big radiation hit. But then this a one-way mission isn't it? Of course, Mars One isn't going to change their mission architecture because of the trifling contamination issue, because viewers want to see boots on the surface. Going round and round in orbit just doesn't have the impact of humans setting foot on the surface of an alien planet. But what about reverse contamination? What if there *is* life on Mars and that life is hazardous to humans? Since it is almost certain that the Mars One contestants will lack immunity to any life that might exist on Mars, chances are they will die. To ensure maximum ratings it would obviously be best if the Martian life killed the contestants slowly rather than immediately. Too low a probability? How do we – or Mars One – know, if that probability hasn't been assessed? Over to you Mars One. And on the subject of contamination, let's not forget to mention dust (Figure 3.7). Remember, back in the Apollo days, the astronauts had all sorts of problems dealing with lunar dust. It got everywhere, including the bronchial system, causing extreme discomfort. Despite the very best seals and hatches, the dust still found a way in and the same scenario is likely to play out on Mars. Now Martian dust has never been studied microscopically and nobody knows if humans are allergic to it. We also don't know if humans are able to build up a resistance to Martian dust and we don't know how damaging it will be when it finds its way into the bronchial tubes. Perhaps the Mars One contestants will suffer continuous hay fever-like symptoms, or perhaps they will suffer a unique type of Martian bronchial pneumonia. What will happen to the reality television show then?

We could on and on about all the problems that contamination and cross-contamination could cause, but there just isn't enough room here. Also, that isn't the purpose of this book: the purpose of this book is to lay open the flaws in the Mars One enterprise and contamination is one of those flaws. You see, spacecraft can be sterilized. Humans can't. What is Mars One's plan for avoiding contamination? No-one knows. Then again, perhaps Mars One doesn't need to follow the planetary protection policies. Not all of them at least. One argument against the very rigorous planetary protection rules is that absolute sterilization (see sidebar) does not exist. Consequently, no matter how many times you treat a spacecraft, that vehicle is still going to carry hundreds of thousands of bugs. The point is we

Figure 3.7   Moon dust. Credit: NASA

probably don't need to sterilize any spacecraft anyway, because deep space radiation does a great job of sterilizing while the vehicle is en route to its destination. In fact, these planetary protection policies regularly get in the way of finding life. Take *Curiosity* for example. This rover has been wandering around Mars for a while now. In 2015, it spotted features known as *recurring slope lineae*, which are thought to be caused by water. Great excitement all around you would have thought. But no. NASA's Office of Planetary Protection stepped in and suggested that *Curiosity* should not live up to its name by investigating the features, because the rover might contaminate the area with terrestrial life that could have hitched a ride. In short, you can apply all the planetary protection measures you want, but the only way you can ensure planetary purity is to conduct exploration through the other end of a telescope. Perhaps the best Mars One can do to keep everyone happy is to offer to stay away from regions of special interest.

Planetary Protection

Preparing spacecraft for a mission to another planet requires a very high level of sanitization. First, there is the precision cleaning that comprises two sterilization steps. The first of these is Dry Heat Microbial Reduction (DHMR), followed by vapor-phase hydrogen peroxide cleaning. To do this, the vehicle is placed in an oven set at 111.7°C and left to bake for 30 hours. After that is done, the vehicle is subjected to more sterilization procedures that must take place in special clean rooms, in which the air is constantly exchanged to prevent contamination.

So what happens if Mars One does contaminate the surface, or if they decide to lay claim to the land on which the habitat is located? Can liability be relevant, and if so by what mechanism? Just how far does the reach of a State actually reach? Perhaps we should let the Martians – because that is probably what the contestants will be referred to – make their own laws? But seriously though, while Mars One might be an ambitious if reckless undertaking, the grand promise of adventure does not mean the organization can simply brush off the legal issues, because without solving these, there is not much of a foundation for success.

4

Selection

Figure 4.0 Public domain

© Springer International Publishing Switzerland 2017

E. Seedhouse, *Mars One*, Springer Praxis Books, DOI 10.1007/978-3-319-44497-0_4

"Men wanted for hazardous journey. Small wages, bitter cold, long months of complete darkness, constant danger, safe return doubtful. Honour and recognition in case of success."

> *One of the most famous advertisements ever, posted by Sir Ernest Shackleton*
> *to recruit a crew for his Imperial Trans-Antarctic Expedition.*
> *5000 men applied for 28 positions.*

Figure 4.1　Jeremy Hansen and David St Jacques. Credit: Canadian Space Agency

"The Canadian Space Agency (CSA) is looking for individuals who want to be part of the next generation of space explorers. Two applicants will be selected to join the CSA Astronaut Corps. A pool of qualified candidates will also be created for future needs.

"The CSA is seeking outstanding scientists, engineers and/or medical doctors with a wide variety of backgrounds. Creativity, diversity, teamwork, and a probing mind are qualities required to join the CSA's Astronaut Corps. To withstand the physical demands of training and space flight, candidates must also demonstrate a high level of fitness and a clean bill of health."

> *CSA advertisement posted May 22, 2008, to recruit two astronauts.*
> *5351 hopefuls applied.*

REAL ASTRONAUT CANDIDATES

Being selected as an astronaut is probably the toughest career goal you can have. Take Rolf, for example. Rolf and I worked at the German Space Agency in the late 1990s while studying for our PhDs. At the time, we had wanted to be astronauts for as long as we could remember and had spent the best part of 20 years gradually accumulating the shopping list of qualifications, just so we could submit a competitive application to our relevant space agencies when the time came. Rolf's final selection came in 2004, by which time he was phenomenally qualified. He had his PhD, a commercial pilot's license and a multi-engine license, he was a certified aircraft mechanic, spoke fluent German, English *and* Russian, was a qualified flight controller for spacewalks, he had hours of zero-G time, and had worked for NASA at Johnson Space Center (JSC) for several years. If I had owned a house at the time, I would have put that house on Rolf being selected. Not just because of his exceptional list of qualifications and not just because he had several senior astronauts backing him at the selection level, but because this guy is everything you imagine an astronaut to be. Absolutely and resolutely committed to manned spaceflight. That's Rolf. He was interviewed but wasn't selected. Why? Well, those of you who have had experience of the astronaut selection business will know how political it is, but that's material for another book. The point is that even someone with Rolf's experience and qualifications didn't get the nod. So how do the Mars One group measure up? Before we answer that let's take a look at someone who did qualify as an astronaut: David St. Jacques (Figure 4.2).

Personal Data: Born January 6, 1970, in Quebec City, Canada, and raised in Saint-Lambert near Montreal, Canada. He is married and has two children. He is a lifelong mountaineer, cyclist, skier and passionate sailor. He also holds a commercial pilot license. David is fluent in English and French and has basic knowledge of Russian, Spanish and Japanese.

Education: David earned a Bachelor of Engineering degree in Engineering Physics from École Polytechnique de Montréal, Canada (1993). He earned a PhD in Astrophysics from Cambridge University, UK (1998), where his

Figure 4.2 David St Jacques. Credit: Canadian Space Agency

studies included theoretical work on astronomical observation and the design, fabrication and commissioning of instruments for the Cambridge Optical Aperture Synthesis Telescope and for the William Herschel Telescope in the Canary Islands. He earned his M.D. from Université Laval in Quebec City, Canada (2005), and completed his family medicine residency at McGill University in Montreal, Canada (2007), where his training focused on first-line, isolated medical practice.

Organizations: Collège des médecins du Québec, College of Family Physicians of Canada, Ordre des ingénieurs du Québec, International Society for Optical Engineering, Life Fellow of the Cambridge Philosophical Society.

Special Honors: Canada Millennium Scholarship (2001 to 2005), Japan Society for the Promotion of Science Post-Doctoral Fellowship (1999 to 2001), Natural Sciences and Engineering Research Council of Canada "1967" Science and Engineering Scholarship (1994 to 1998), Canadian Space Agency Supplement Scholarship (1994 to 1998), Cambridge Commonwealth Trust Honorary Scholar (1994 to 1998), United Kingdom Overseas Research Student Award (1994 to 1998), Canada Scholarship (1989 to 1993).

Experience: Prior to joining the Canadian Space Program, David was a medical doctor and the Co-Chief of Medicine at Inuulitsivik Health Center in Puvirnituq, Canada, an Arctic village on Hudson Bay. He also worked as a Clinical Faculty Lecturer for McGill University's Faculty of Medicine, supervising medical trainees in Nunavik. David began his career as a biomedical engineer, working on the design of radiological equipment for angiography. His broad scientific background also includes astrophysics and medical training. His post-doctoral research included the development and application of the Mitaka Infrared Interferometer in Japan and the Subaru Telescope Adaptive Optics System in Hawaii (1999 to 2001), after which he joined the Astrophysics group at Université de Montréal. His international experience also includes engineering study and work in France and Hungary and medical training in Lebanon and Guatemala.

David was one of the candidates in the CSA's Astronaut Recruitment Campaign that I was part of. Anyone with a PhD in astrophysics (from Cambridge University to boot!) *and* a medical degree was always going to be a strong candidate, so it was no surprise when he was selected along with Jeremy Hansen, a Canadian Forces fighter pilot. And look at all of David's other qualifications! Outstanding. Just outstanding. How do the Mars One candidates measure up? Well, let's see shall we. Hmmm, so many to choose from, but let's start with Joanna Hindle, a contestant who made it into the final 100. Joanna is a Canadian in her 40s who lives in Whistler, where she works for the Lil'wat Nation as an English and social studies teacher at a community school. I'm sure those will be useful skills en route to Mars! Our Canadian Mars One wannabe lists her interests as learning, reading, pondering, writing, dreaming, outdoorsing, and laughing. I swear I'm not making this up. Astronaut material? Take a look at David's resume again before answering that question. The qualifications of the rest of the Mars One group hardly set the world on fire, although there is the odd pilot and medical doctor. But before we take a closer look at this new breed of reality TV spacefarer, I think it's useful to give you an idea of what this astronaut business is all about (Figure 4.3), because it seems the Mars One organization doesn't have a clue. Mars One selectors take note.

Figure 4.3   Julie Payette. Credit: Canadian Space Agency

SELECTION FOR A SERIOUSLY RISKY JOB

To begin with, there is no occupation that is so far from ordinary than being an astronaut. Or more risky. Consider this statistic: of the 543 people who have reached Earth orbit as of January 2016, 18 have died, which equates to a mortality rate of 3.74%. This makes being an astronaut one of the most lethal professions you could contemplate. In fact, it is the risk equivalent of 400 Boeing 747s crashing *every* day, or 197,000 passengers being killed every day. So why do people apply to be astronauts? Well, in my experience, those who submit to the year-long application process do so because of a love for space and a burning and overwhelming desire to contribute to a noble and grand endeavor. Aligned with these powerful drivers is the desire to work with exceptional people who share the same vision. To this very elite bunch, the risks and the long hours are deemed minor inconveniences. And because of these potent motivations, space agencies have never, *ever* had a problem filling slots in the astronaut ranks (Table 4.1).

Table 4.1   NASA Astronaut Recruitment

Year	Applications	Selected	%	Year	Applications	Selected	%
1978	8,000	35	0.004	1995	3,100	19	0.006
1980	3,500	19	0.005	1996	2,600	35	0.01
1984	4,900	17	0.003	1998	2,600	25	0.009
1987	2,100	15	0.007	2000	3,000	17	0.005
1990	2,400	23	0.009	2004	2,900	11	0.003
1992	2,300	19	0.008	2008	2,900	15	0.005

Those very elite few who are serious about applying for the planet's most demanding job have invariably spent the best part of two decades accumulating the qualifications necessary to submit a competitive application. People like Rolf. Or David. Or Jeremy. All those who make it to the interview stage have a stellar educational background. Stellar. Sure, the minimum requirements to apply include a bachelor's degree (see Panel 4.1 for an example), but no selection committee is going to give your application a second look unless you have a PhD, combined with an outstanding professional background and the requisite myriad astronaut-specific qualifications – diving, pilot, sky-diving: take your pick – thrown in.

And having a stunning set of qualifications is just for starters, because you also need to be supremely healthy. With the dawn of long duration missions on board the ISS, the medical standards have become more stringent than ever. That's because the statistical risk for a medical event increases with mission duration. For example, an episode of renal colic in a mission specialist may have been acceptable for a short duration Space Shuttle

Panel 4.1. CSA General Selection Requirements

The applicant must be a Canadian citizen or a resident of Canada.

Because astronauts are required to perform a broad range of scientific and technical work, prospective candidates must hold a bachelor's degree, recognized in Canada, in one of the following areas:

- Engineering or Applied Sciences
- Science

The bachelor's degree must be followed by at least two years of related professional experience.

OR:

A bachelor's degree along with a master's degree or a doctoral degree, recognized in Canada, in one of the following areas:

- Engineering or Applied Sciences
- Science

OR:

- A license to practice medicine in a province or a territory of Canada.

Medical Requirements

To be selected, applicants must meet stringent medical criteria. Applicants will be required to undergo Canadian Space Agency medical physical exams, which include the following specific requirements:

- Standing height must be between 149.5 cm and 190.5 cm.
- Visual acuity must be 20/20 (6/6) or better in each eye, with or without correction.
- Maximum limits for cycloplegic refractive error and astigmatism correction apply.

  The refractive corrective surgical procedures PRK or LASIK are allowed. For those candidates under final consideration, an operative report on the surgical procedure will be requested. The CSA does not recommend that potential candidates undergo laser refractive surgery for the sole purpose of applying for employment as an astronaut.

- Blood pressure must not exceed 140/90 mm Hg, measured in a sitting position.
- Meet the following pure tone audiometry hearing thresholds:

Frequency (Hertz):   500 1000 2000 3000 4000
Either ear (decibels): 30   25   25   35   50

Conditions of employment

- The position will require frequent travel and relocation.
- The position is subject to pre-employment security clearance.
- Candidates must undergo pre-employment medical examinations.

mission, but is disqualifying for an Exploration-Class mission. This is because pre-flight ultrasound screening can rule out significant retained calculi and the probability of developing a calculus during a 10- to 14-day Shuttle mission is very low. While most people assume they are healthy, it is worth highlighting some statistics from the CSA's 1992 astronaut recruitment campaign (see Table 4.2), which sent out 337 questionnaires to the most qualified applicants.

Table 4.2 Medical disqualification during CSA's 1992 Astronaut Selection

Cause for Disqualification	Number of Applicants Disqualified	% of Applicants Disqualified (N = 23)
Cardiovascular	14	61%
Vision	7	30%
Thyroid	2	9%
Ear Nose Throat	1	4%
Respiratory	1	4%
Gastrointestinal	1	4%
Genitourinary	1	4%
Neurological	1	4%

So you are medically qualified, you have an outstanding resume and you have the necessary avocational skills (Figure 4.4) such as a private pilot's license, parachuting experience, and a scuba-diving license. But what about the *qualities* that make you an astronaut candidate? Well, one virtue you will need is *patience*. Bucket-loads of it. This will apply to the Mars One hopefuls as much as it applies to the current crop of astronaut candidates, because astronauts must be prepared to spend years and *years* training for their mission. According to the Mars One website, by the time launch day finally rolls around, the crew will have waited the best part of a decade for their flight.

> "A human being should be able to change a diaper, plan an invasion, butcher a hog, conn a ship, design a building, write a sonnet, balance accounts, build a wall, set a bone, comfort the dying, take orders, give orders, cooperate, act alone, solve equations, analyze a new problem, pitch manure, program a computer, cook a tasty meal, fight efficiently, die gallantly. Specialization is for insects."
>
> *Robert A. Heinlein.* 'Time Enough for Love'

The great sci-fi author is absolutely right. Mars One candidates had better be a versatile lot. And by versatile, I mean they should have high technical competence and the ability to work in a team. In addition to having the patience of a saint and being highly versatile, the Mars One candidates should all be high achievers. It's all well and good having a PhD and a flying license, but astronauts also tend to have some noteworthy accomplishment outside their professional field of expertise. Perhaps in addition to being a physician you've climbed Mount Everest, or perhaps you're a fighter pilot in addition to having raced in the

Figure 4.4 Analog training outside the NEEMO habitat. Credit NASA

Olympics (I have colleagues who have done both). How many in the Mars One 'Final 100' fit the bill of high achievers? Not many, unfortunately. Unless you count 'laughing' as an accomplishment I guess! Oh dear.

> "Provided you select the right individuals and intervene before serious problems, there doesn't seem to be a reason why humans couldn't handle going to Mars, even forever."
>
> *From* 'Leaving Earth. Why One-way to Mars Makes Sense'.
> *Andrew Rader. BISAC.*

Now let's move on to psychological disposition. This is very important because our Mars One crew will spend the rest of their albeit very, *very* short lives in a confined space with three other contestants. It goes without saying that the psychological qualities required will include an ability to get on well with others, an affinity for team work and adaptability. Those selected will have to be able to adapt quickly to changing situations and exercise mature judgment, just two of the key qualities they are going to find helpful in performing tasks in an austere environment. Another component of psychological disposition is public relations: in common with their government-employed counterparts, Mars One contestants will spend an awful lot of time doing PR work. The role of a Mars

One contestant as a PR officer will be a natural one, since the media and the public will be curious about the mission. This means our Mars One contestants must enjoy meeting the public and the press, and be able to communicate the importance of their tasks in space. And since these contestants will be the public face of Mars One, personality will inevitably be a significant selection criterion when it comes down to deciding on the final four for the first mission. So exactly how will Mars One select this intrepid four? We'll get to that, but before we do it is instructive to delve into how the agencies select their astronauts. After all, they've been doing this for some time now, so perhaps the Mars One organization can take a few pointers.

To begin with, not all astronaut selection procedures are alike. While space agencies[1] search for only the best of the very best, selection methods differ. NASA, for example, conducts a panel interview, together with routine fitness, medical and coordination tests, which is an approach similar to the selection performed by ESA. In contrast, the CSA, which has by far the most rigorous selection process of any agency in the world, requires applicants to submit to myriad tests, many of which wouldn't be out of place in military boot camp.

In NASA's 2008 selection, the agency received 3535 applications. Following a thorough review of the information, 120 hopefuls were invited to visit JSC in October 2008, where the first item on the assessment agenda was a welcome by the selection office. This was usually conducted by the chair and deputy chair of the astronaut selection board, in this case astronauts Peggy Whitson and Steven Lindsay. They explained to the group what it was like to be an astronaut and what the chances of an early exit were! After a series of anthropometric measurements, to make sure each candidate would fit into the Soyuz capsule, there followed a robotics evaluation that determined 3D reasoning skills, situational awareness, and the ability to multitask. Next was a series of written psychological exams, requiring applicants to answer thousands of questions to assess sociability, teamwork, and all sorts of other psychological parameters. The process concluded with a one-hour panel interview administered by the selection board. To introduce themselves to the 12-person board, each applicant wrote down three to five reasons why they wanted to be an astronaut (the Canadian applicants, in contrast, had to write a 1000-word essay!). In March 2009, 40 candidates were chosen for the next phase of selection, comprising a week-long medical. Finally, on June 29, 2009, NASA selected nine astronaut candidates[2] (Table 4.3).

---

[1] In 2008, NASA, the Canadian Space Agency (CSA) and the European Space Agency (ESA) all initiated astronaut selection campaigns. The odds were daunting. NASA received 3535 applications for 15 positions, which meant prospective astronauts had a 0.4% chance of being selected (235 people chasing each position). ESA received 8413 applications for 4 positions, meaning future European spacefarers had only a 0.04% chance of being selected (2103 people chasing each position). The highest odds? The poor Canadians, whose agency received 5352 applications for just 2 positions (2676 people chasing each astronaut position!).

[2] Three weeks later the agency announced an additional seven candidates (Mike Schmidt, Stephen Heck, Stuart Witt, Jim Kuhl, Lanette Oliver, Chantelle Rose, Rachael Manzer, and Maureen Adams), as part of the Educator astronaut selection.

Table 4.3 NASA's Class of 2009

Name	Age	Background
Serena M. Aunon	33	Wyle flight surgeon for NASA's Space Shuttle, ISS and Constellation Programs. Holds degrees from The George Washington University, University of Texas Health Sciences Center in Houston, and UTMB.
Jeanette J. Epps	38	Technical intelligence officer with the CIA. Holds degrees from LeMoyne College and the University of Maryland.
Jack D. Fischer	35	Test pilot and U.S. Air Force Strategic Policy intern at the Pentagon. Graduate of the U.S. Air Force Academy and Massachusetts Institute of Technology (MIT).
Michael S. Hopkins	40	Lt. Colonel U.S. Air Force. Special assistant to the Vice Chairman (Joint Chiefs of Staff) at the Pentagon. Holds degrees from the University of Illinois and Stanford University.
Kjell N. Lindgren	36	Wyle flight surgeon for NASA's Space Shuttle, ISS and Constellation Programs. Degrees from the U.S. Air Force Academy, Colorado State University, University of Colorado, the University of Minnesota, and UTMB.
Kathleen Rubins	30	Principal investigator and fellow, Whitehead Institute for Biomedical Research at MIT and conducts research trips to the Congo. Degrees from the University of California-San Diego and Stanford University.
Scott D. Tingle	43	Commander U.S. Navy. Test pilot and Assistant Program Manager-Systems Engineering at Naval Air Station Patuxent River. Degrees from Southeastern Massachusetts University and Purdue University.
Mark T. Vande Hei	42	Lt. Colonel U.S. Army, Flight controller for the ISS at NASA's Johnson Space Center, as part of U.S. Army NASA Detachment. Graduate of Saint John's University and Stanford University.
Gregory R. Wiseman	33	Lt. Commander U.S. Navy. Test pilot. Department Head, Strike Fighter Squadron 103, USS Dwight D. Eisenhower. Graduate of Rensselaer Polytechnic Institute and Johns Hopkins University.

Meanwhile, on the other side of the pond, prospective ESA candidates busied themselves being tested for the JAR-FCL 3, Class 2 medical examination, since it was this that comprised the first step in the European selection process. Following the first round of evaluation, ESA selected 918 candidates for computer-based psychological testing. From this first batch, ESA then invited 192 candidates to the second stage of psychological testing, which took place at the European Astronaut Center (EAC), Cologne, Germany. From these 192 hopefuls, ESA selected just 80 to return for extensive medical evaluation and following this stage, 40 candidates were invited for a formal panel interview, from which 6 candidates (Table 4.4) were selected (Figure 4.5).

Figure 4.5 The European Space Agency astronauts selected in the 2008-2009 campaign. Credit: ESA

Table 4.4   ESA's Astronaut Selection 2008

Name	Age	Background
Samantha Cristoforetti	32	Master's degrees in engineering and in aeronautical sciences from the University of Naples Federico II in Italy. Worked as a fighter pilot with the Italian Air Force.
Alexander Gerst	33	Diploma in geophysics. Studied Earth science at Victoria University of Wellington in New Zealand, where he was awarded a Master of Science. Working as a researcher since 2001.
Andreas Mogensen	33	Master's degree in engineering from Imperial College, London and a doctorate in engineering from the University of Texas. Worked as an attitude & orbit control system and guidance, navigation & control engineer for HE Space Operations.
Luca Parmitano	33	Holds a diploma in aeronautical sciences from the Italian Air Force Academy. Trained as a Full Experimental Test Pilot at EPNER, the French test pilot school in Istres. Pilot with the Italian Air Force.
Timothy Peake	37	Degree in flight dynamics and qualified as a Full Experimental Test Pilot at the UK's Empire Test Pilots' School. Officer serving with Her Majesty's Forces as an Experimental Test Pilot.
Thomas Pesquet	31	Master's degree from the Ecole Nationale Supérieure de l'Aéronautique et de l'Espace in Toulouse, France. Worked at the French space agency, CNES, as a research engineer. Currently flies Airbus A320s for Air France.

In tandem with NASA and ESA, the CSA was busy with its third astronaut recruitment campaign, which was by far the most daunting of any astronaut selection to date. It all started in St. Jean, home to the Canadian Forces (CF) basic training base, where candidates were subjected to myriad exercise tests and assessments, beginning with the CF's basic fitness test with a few twists. The first test was the Multi-Stage Fitness Test (MSFT),[3] aka 'the bleep test', familiar to everyone with a military background. After the bleep test (Figure 4.6), it was time for press-ups, sit-ups, chin-ups, and some fiendishly uncomfortable isometric tests.

Figure 4.6 Multi-stage fitness test. Credit: Canadian Space Agency

Next was a series of swimming and sea survival tests. On the initial application form, there had been a question that asked if the applicant was comfortable with water. I would imagine most people would answer that in the affirmative, since you have to drink it every

---

[3]The MSFT estimates maximum oxygen uptake. The test involves running continuously between two points 20 meters apart. These shuttle runs are synchronized with an audio tape, which beeps at set intervals. As the test proceeds, the interval between each successive beep reduces, forcing the victim to increase speed until he/she is incapable of keeping up. The recording is structured into 23 'levels', each lasting around 63 seconds. The interval of beeps is calculated to require a speed at the start of 8.0 km/h, increasing by 0.5 km/h with each level. The progression from one level to the next is signaled by 3 rapid beeps. The highest level attained before failing to keep up is recorded as the score for that test.

day to stay alive! What the question *should* have asked was whether the candidate could swim! The first swim test was a very leisurely 250 meters in 10 minutes. This was followed by a dive into the pool from three meters and treading water for 10 minutes with hands held above the surface. Candidates then jumped into the water from the 5-meter diving board and swam to the side wearing a life preserver, before conducting the underwater 'shapes' task which required each candidate to dive to the bottom of the pool and fit different shaped blocks of wood into a container.

The second day of testing was conducted at the CSA's Saint Hubert location, where candidates completed the CF's aircrew selection test, together with a series of more than 20 computer-based tests designed to assess cognitive, spatial and motor skills. Next up was a series of hand-eye coordination tests on paper, followed by a public service exam, usually written by diplomats. In between testing sessions, candidates would be pulled aside to conduct an impromptu media interview in front of television cameras. After the public service exam, candidates were served 1200 questions in three psychology questionnaires, at the end of which everyone's hands were cramping badly. The third day of testing focused primarily on robotics (Figure 4.7). Unlike the two-week course offered to actual astronauts during their advanced and increment-specific training, candidates received just two hours of instruction on how to manipulate the Canadarm2. The third day's program brought to a close this phase of testing. Next on the path to being a CSA astronaut was the Navy's Damage Control School in Halifax.

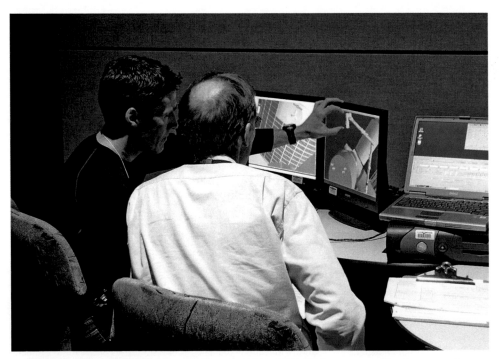

Figure 4.7   Robotics assessment during the CSA's 2008-2009 astronaut selection campaign. Credit: Canadian Space Agency

A month after Saint Hubert, 31 candidates found themselves in Halifax at the Navy's Damage Control School, a high-fidelity mock-up of a warship. Inside the Damage Control School, sailors are taught how to fight fires and flood, in addition to learning how to react to hazardous material (HazMat) spills. After suiting up in coveralls and wetsuits, candidates entered a flood compartment that slowly filled with icy water. Peering through observation windows were CF damage control instructors and members of the CSA's astronaut selection group, who made notes as to how well (or badly!) the candidates worked as team in a stressful environment. After fighting floods, the candidates were rushed off to don bunker gear so they could fight fires. Dressed in their bunker gear and holding a fire hose, teams were led into a fire trench, where they put their limited fire-fighting training to the test under the watchful eyes of the selection staff (Figure 4.8).

The final day of testing was held at Survival Systems. Candidates were taken one at a time to the helo-dunker (Figure 4.9), where they were strapped into a mock helicopter cockpit and given escape instructions before being sent crashing into the swimming pool. The dunker then rolled over and quickly filled with water. An instructor then tapped each candidate on the shoulder, which was the signal for escape. After surviving the dunking, candidates were led to the top of a ten-meter tower and told to jump into the swimming pool and clamber into a life-raft.

Figure 4.8 Fire fighting assessment during the CSA's 2008-2009 astronaut selection campaign. Credit: Canadian Space Agency

Figure 4.9 Dunker training during the CSA's 2008-2009 astronaut selection campaign. Credit: Canadian Space Agency

After being exposed to fire, flood *and* hypothermia, the candidates returned home and waited for the email informing them whether or not they had progressed to the final 16. The final cut to 16 eliminated some extraordinarily well-qualified candidates, many of whom would probably have been selected as astronauts if only they had had an American passport. In fact, a couple of those working in the United States who missed the final cut were already making plans to obtain a Green Card, so they would be eligible for NASA's next selection. The final 16 were introduced at a press conference at the Defense Research and Development facility in Toronto in March. And, after a job interview lasting the best part of a year, two candidates finally made the grade: Dr. David Saint-Jacques and Captain Jeremy Hansen (Table 4.5).

Table 4.5 CSA's Successful Astronaut Candidates 2009

Name	Age	Background
Jeremy Hansen	33	Bachelor of Science degree in Space Science from Royal Military College in Kingston, Ontario. Master of Science in Physics from the same institution in 2000. Prior to joining the Canadian Space Program, Jeremy served as a CF-18 fighter pilot and held the position of Combat Operations Officer at 4 Wing Operations in Cold Lake, Alberta.
David Saint-Jacques	39	Bachelor of Engineering degree in Engineering Physics from École Polytechnique de Montréal. PhD in Astrophysics from Cambridge University, UK. MD from Université Laval in Quebec City, Quebec. Completed family medicine residency at McGill University. Prior to joining the Canadian Space Program, David was a medical doctor and Co-chief of Medicine at Inuulitsivik Health Centre in Puvirnituq, Quebec.

MARS ONE SELECTION

"Mars One will carefully select the crew for a number of skills and qualities. They will be people who have dreamt their whole life of going to Mars, and in many cases will have pursued careers that increase the odds of being selected for this kind of mission. The selected team will be very smart, skilled, mentally stable and very healthy. They will go to Mars to live their dream."

Mars One website blurb

So now you have an insight into how governments select astronauts. How will Mars One go about the process? Well, not selecting those responsible for headlines like this would be a start:

'Adam and Eve of another planet':
Briton bids to be first woman to give birth to a MARTIAN.
WOULD you Adam and Eve it?

"A British student is in line to become the first woman ever to give birth to a MARTIAN. Maggie Lieu is hoping to become the first woman to give birth on Mars. Lieu, 24, has applied to be part of the first human colony on the red planet through the £3.9 billion Mars One Project. It aims to create a permanent settlement for our race on Mars and will start by sending 40 people in 2024 after 10 years of extensive training. She has been shortlisted to make the 140 million mile one-way journey to Mars and she believes she could be the first woman to give birth on Mars. Miss Lieu said: "It would be incredible to be the Adam and Eve of another planet. Because it is a colonisation programme, it's inevitable that eventually someone will procreate and it would be incredible to be the first mother on Mars."

Article written by Dion Dassanayake, January 30, 2015

The worrying thing about the 'Adam and Eve...' headline is that it was generated by an applicant Mars One saw fit to invite to the second round of selection. Scary! Clearly the organization has little knowledge of this astronaut selection business. So what's next? Well, as we close in on the autumn of 2016, Mars One has selected the Mars 100.[4] This group is now in the third round of the selection. During this round, the applicants will group together in teams of 10 to 15 members and be set group-dynamic challenges; how to deal with a contestant who wants to give birth on the Red Planet might be one scenario perhaps? After five days of problem-solving in their teams, the Mars One selection committee – none of whom have been an astronaut, incidentally – will select 40 candidates for

---

[4] 50 men and 50 women representing countries from around the world, including 42 from the Americas, 28 from Europe, 16 from Asia, 7 from Africa, and 7 from Oceania.

the Isolation Challenge. This ominous sounding task is actually just a nine-day isolation exercise that supposedly decides which contestants will work best together. Nine days! After those nine days, another ten contestants will be cut. The final 30 will then be subjected to the Mars Settler Suitability Interview that will assess the contestant's suitability for a long-duration space mission. Really? The final 30 will then be whittled down to 24 and it will be this group that starts the training.

> "The large cut in candidates is an important step towards finding out who has the right stuff to go to Mars. These aspiring Martians provide the world with a glimpse into who the modern day explorers will be."
>
> *Mars One co-founder and CEO Bas Lansdorp*

So how did Mars One select its third round group? Well, the candidate screening process was developed by the organization's Chief Medical Officer, Norbert Kraft, who once upon a time worked for NASA. Of the 4,227 completed applications from the first round, 1,058 were selected for the second round based on a questionnaire that asked applicants how they might handle stressful situations. Selection to the third round was based on submission of a medical exam, similar to the one required by NASA, which reduced the number of eligible candidates to 660. These 660 were interviewed by Dr. Kraft, who asked the second round contestants questions about Mars and life support. The interview lasted about 10 minutes. Based on the results of the interviews, the Mars One organization selected the Mars One 100 hopefuls in February 2015. Not long after, headlines like this began appearing in the internet ether:

> "Shocker: Mars One mission begins to unravel as finalist slams selection process."

This particular header appeared in March 2015 and the subject of the article was a disgruntled Mars One candidate, Dr. Joseph Roche, who argued that the Mars One enterprise was nothing more than a poorly conceived and even more poorly executed mission that was going nowhere. Dr. Roche, a professor of physics and astrophysics (over-qualified for Mars One in other words) at the School of Education, Trinity College, Dublin, went on to argue that the science of Mars One was shaky at best and that contestants would probably not be provided with adequate radiation protection. Dr. Roche concluded by saying he didn't think he would see a one-way mission in his lifetime. Dr. Roche's comments prompted a flurry of activity by the Mars One bloggers, who denounced the professor's heretical comments. Some pointed to the upcoming (at the time) one-year mission by Scott Kelly and Mikhail Kornienko, arguing that their mission would lead to a better understanding of the challenges faced by astronauts embarking on long-duration flights, before somehow connecting the dots to make an argument for how this mission would lead to Mars One succeeding. Crazy! But who are the other 99 candidates? Well, I've listed half a dozen below. Read their introductions and then compare the qualities of these candidates with those of David and Jeremy, and of someone who has already flown in space, ESA astronaut Samantha Cristoforetti. Then ask yourself this question: does Mars One have even the vaguest idea of how to select astronauts?

SELECTED CANDIDATES FROM THE MARS ONE SHORTLIST

Christian O. Knudsen
Age: 34
From: Denmark

Christian says:
"I believe the potential benefits of the Mars One project far outweigh the potential costs it may have to me, personally. I believe these benefits will be scientific progress, which can benefit all of us on Earth. If you compare the Mars One mission to the moon landing, I think scientific progress, on a similar scale to what we experienced following that endeavour, is a reasonable expectation. Another benefit of the Mars One project, in my mind, is the motivation it ignites in other people, the surge in students choosing an education in the fields of science and engineering following the Apollo space programme is, in my opinion, a result of this motivation.

"Furthermore, personally, and without any scientific backing, I believe that the increase in living standard these advances allow will leave more space for individuals to expand the sphere of people they care for and will sacrifice for, beyond themselves, beyond their family and beyond their nationality. As idealistic and altruistic as all this may sound, I'm also personally motivated by the desire to test limits, personal as well as technological."

Josh Richards
Age: 29
From: Australia

Josh says:
"Hi there! My name is Josh and I'm a physicist/engineer turned stand-up comedian. I've been lucky enough to travel, work & learn in the military, sciences & arts; & I love sharing these experiences onstage – all with a huge grin on my face! I knew I wanted to explore space when I was 10 and saw Australian astronaut Andy Thomas on TV during STS-77. My path here has been a wandering one, but I dream of using my diverse background and problem-solving skills to support Mars One's incredible mission."

Etsuko Shimabukuro
Age: 50
From: Japan

Etsuko says:
"The First Sushi Bar on Mars. The girl had a telescope and dreamed of living on another planet someday. As a young woman she studied archaeology to understand where she came from. She explored limestone caves to feel the million years of planet Earth. She studied computer science because she was fascinated by the potentiality of the virtual world.

"Then she went into the IT industry to work with people blessed with incredible intelligence. In her thirties she backpacked for two years around the world to understand humanity and life on Earth. She also walked 7,500km (4,600 miles) through Japan to be a little help for charity and to understand every corner of her homeland. In her forties she ran for seven days in the Sahara Desert in Morocco to test her mental and physical boundaries.

"She became an ascetic mountain priest in order to live simply. Later, she studied Japanese cuisine to make the world healthy. She presently lives and works in Mexico as a Japanese chef to share the wisdom of her ancestors.

"She is very grateful to have had so many opportunities to fulfill her life on Earth. Now she plans to go to Mars to be at the forefront of human evolution and to open the first sushi bar on Mars. Her life is one big experiment."

Dianne McGrath
Age: 45
From: Australia

Dianne says:
"Hi! I am Di! I am a leader, team player, project manager and creative problem-solver. My experience working in and managing teams in sales, marketing, emergency services, government and the not-for-profit sectors provides me with a strong platform to work as a unit to deliver on outcomes.

"My sense of adventure, determination and fitness has seen me sail tall ships in the Southern Ocean, cycle extreme distances, jump from planes, and run marathons and an ultramarathon. I believe nothing is insurmountable.

"I have a special interest and growing expertise in sustainable food systems – something that will be critical in a completely new environment for human habitation. My current postgraduate study is focused on food waste and sustainable food systems, and involves the research project *Watch My Waste*."

Steve Schild
Age: 30
From: Switzerland

Steve says:
"I am a young, motivated, fun-loving person. I love meeting new people. I am full of ideas and boredom is not something that I am familiar with. I have a very inquisitive mind and am always looking for new challenges. I care about my well-being and that of the environment. My friends say that it is impossible for me to sit still for a minute, as I always have to be doing something! For me, there are no setbacks in life, just opportunities to learn something new. My motto is: always continue to look forward."

Lennart Lopin
Age: 35
From: Germany

Lennart says:
"I am a 35-year-old software developer from Florida, USA. As a former Buddhist monk I spent a couple of years meditating in caves and probably know a thing or two about sensory deprivation ;-). I am very enthusiastic about this project and a big fan of Zubrin's *A Case for Mars*. IMHO, the future of our civilisation depends on our ability to live independently of planet Earth and to spread into the solar system. It would be an honor to work towards that goal with my fellow Martian colonists."

A REAL ASTRONAUT: SAMANTHA CRISTOFORETTI

Personal data: Born in Milan, Italy, on April 26, 1977, Samantha Cristoforetti enjoys hiking, scuba diving, yoga, reading and traveling. Other interests include technology, nutrition and the Chinese language.

Education: Samantha completed her secondary education at the Liceo Scientifico in Trento, Italy, in 1996, after having spent a year as an exchange student in the United States.

In 2001, she graduated from the Technische Universität Munich, Germany,

Figure 4.10   Samantha Cristoforetti. Credit: ESA

with a master's degree in mechanical engineering, with specializations in aerospace propulsion and lightweight structures. As part of her studies, she spent four months at the Ecole Nationale Supérieure de l'Aéronautique et de l'Espace in Toulouse, France, working on an experimental project in aerodynamics. She wrote her master's thesis in solid rocket propellants during a 10-month research stay at the Mendeleev University of Chemical Technologies in Moscow, Russia.

As part of her training at the Italian Air Force Academy, she also completed a bachelor's degree in aeronautical sciences at the University of Naples Federico II, Italy, in 2005.

Experience: In 2001, Samantha joined the Italian Air Force Academy in Pozzuoli, Italy, graduating in 2005. She served as class leader and was awarded the Honor Sword for best academic achievement. From 2005 to 2006, she was based at Sheppard Air Force Base in Texas, USA. After completing the Euro-NATO Joint Jet Pilot Training, she became a fighter pilot and was assigned to the 132nd Squadron, 51st Bomber Wing, based in Istrana, Italy.

In 2007, Samantha completed Introduction to Fighter Fundamentals training. From 2007 to 2008, she flew the MB-339 and served in the Plan and Operations Section for the 51st Bomber Wing in Istrana, Italy.

In 2008, she joined the 101st Squadron, 32nd Bomber Wing, based at Foggia, Italy, where she completed operational conversion training for the AM-X ground attack fighter.

Samantha is a Captain in the Italian Air Force. She has logged over 500 hours flying six types of military aircraft: SF-260, T-37, T-38, MB-339A, MB-339CD and AM-X.

Samantha was selected as an ESA astronaut in May 2009. She joined ESA in September 2009 and completed basic astronaut training in November 2010. In July 2012, she was assigned to an Italian Space Agency ASI mission aboard the International Space Station. She was launched on a Soyuz spacecraft from Baikonur Cosmodrome in Kazakhstan on November 23, 2014, on the second long-duration ASI mission and the eighth long-duration mission for an ESA astronaut.

Samantha worked and lived on the International Space Station for almost 200 days as part of her Futura mission and enjoys interacting with space enthusiasts on Twitter as @AstroSamantha.

Samantha's bio is taken from the ESA website: http://www.esa.int/Our_Activities/ Human_Spaceflight/Astronauts/Samantha_Cristoforetti

Think I'm being too harsh? Let's read a snippet offered by Dr. Kraft on the subject of how Mars One will go about selecting the final 40.

"I can't reveal the specific challenges, since the selection process involves candidates learning to solve problems as a team, but as an example, the Mars One selection committee might blindfold a group of people and tell the group that the goal is to make a perfect triangle. The Mars One selection committee then observes how the candidates solve the problem as a team. How did candidates decide to proceed? How did candidates organize themselves into a team? How did candidates handle the conflicts that inevitably emerge while facing the challenge?

"This process will play out over five days, and from our observations we will be able to screen the Mars 100 to 40 potential candidates. These candidates will begin the isolation part of the screening process, which will take place over another nine days. On a long voyage and in a permanent settlement, people in a small group can't hide or avoid each other. They will have 24 hours a day to annoy each other! This means simple things matter quite a bit – for example, whether you could be very bothered by dirty socks on the floor, dirty dishes in the sink, or body odor."

Dirty socks and body odor! Five days to decide who will go to the final stage of selection? Blindfolding? Give me a break. Look, the astronauts I know were all certain they wanted to be astronauts from a very, very early age, and they pursued that goal with relentless persistence and determination for decades. Some of them went through two, three or even four astronaut selections before they were finally selected. How many of the Mars One candidates show any evidence of having this commitment to becoming an astronaut? How many of them have even been through the first couple of stages of an astronaut selection? Exactly. It's all well and good to spout about how committed you are to becoming an astronaut, but if you're going to go blathering on the internet about it, you'd better back up your statements with some hard evidence and I just don't see it in any of the Mars One candidates. Even a decade or more of training won't dial up the commitment to the level needed to turn Mars One candidates into astronauts capable of surviving even the first month of the trip, never mind achieving the mission goal. That's because that level of commitment – the commitment to pursue a goal over two decades – is a very, very rare quality indeed. Astronauts have it in spades. Take another look at Samantha Cristoforetti's resume and then compare her list of qualifications to those posted by the Mars One group. Who would you select?

5

Training

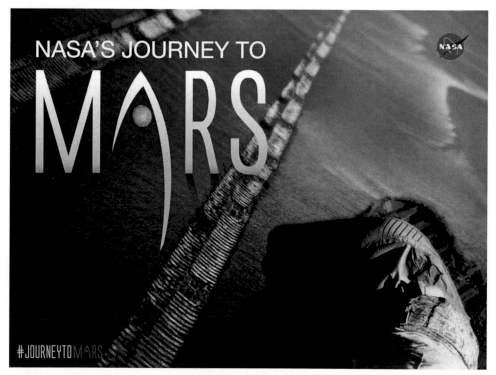

Figure 5.0   Credit: NASA

© Springer International Publishing Switzerland 2017
E. Seedhouse, *Mars One*, Springer Praxis Books, DOI 10.1007/978-3-319-44497-0_5

"There is one Chinese old saying: 'It is not regretful to die in the evening if I can learn the truth in the morning'. Life is short but the universe is ultimate. Human beings are so tiny compared to space. So if tiny me could get closer to the truth of the universe, any cost is insignificant. If this journey does not work, at least I make my effort towards the truth and it will contribute to the final success."

Mars One applicant's metaphysical take
on the one-way to Mars business.

As with so much that is Mars One, there is scant information on the subject of how the reality show will train its contestants, but this is what we do know. Once the candidates have been trimmed to 24 in number, they will begin training in teams of four. This training will take 10 years, which is a little longer than the usual three to four years required of a government employed astronaut to be mission-ready, unless you happen to be Canadian. Training, according to the blurb on the Mars One website and in its book (*Mars One: Humanity's Next Great Adventure*), will comprise a series of 'group dynamic challenges'. One such challenge the Mars One selection committee might spring on the candidates, as already mentioned, is to blindfold a group of people and tell the group that the goal is to make a perfect triangle. That little snippet of information was taken from their website: it's good to know so much thought has gone into the selection and training! For those interested in the details, the following is taken from the Mars One website:

Mars One Astronaut Training Program:

After the crews are selected they begin training. The training consists of three [phases] including technical, personal, and group training.

Phase 1: Technical training includes the training of two crewmembers to be proficient in the use and repair of all equipment to the extent that they can identify and solve technical problems. In addition, two crewmembers will receive extensive medical training in order to treat minor, major, and critical health problems. At least one person will train in the studies of Mars geology and the remaining person will gain expertise in exobiology, which is the biology of alien life.

Phase 2: Personal training consists of ensuring that the astronauts are able to cope with the difficult living environment on Mars. Since these individuals will be unable to speak to friends and family on Earth face-to-face, a certain amount of coping skills are essential.

Phase 3: Group training will mainly take place through simulation missions. During these simulations, the astronauts [will] take part in a fully immersive exercise that prepares them for the real mission to Mars. The simulated environment will invoke as many of the Mars conditions as possible. Immediately after selection, the groups will participate in these simulations for a few months per year.

HOW REAL ASTRONAUTS TRAIN

It all sounds a little vague, doesn't it? And it's not as if the information about how to train astronauts isn't readily available, because governments have been training astronauts for almost 60 years. Here's a snapshot of how the European Space Agency (ESA) goes about training their spacefarers. ESA's astronauts begin their training cycle with the 'Basic Training' phase, which is conducted at the European Astronaut Center (Figure 5.1) in Cologne, Germany. This phase begins with an orientation to the job of being an astronaut. Once that is checked off, the astronauts move on to the fundamentals of the job, a training block that includes basic knowledge of the myriad technical and scientific disciplines – spaceflight engineering, orbital mechanics, space physiology, aerodynamics – that are relevant to being an astronaut. Next up is the 'Space Systems and Operations' training block, which includes an in-depth orientation to the many, many systems of the International Space Station (ISS), and an introduction to guidance, navigation and control, payload systems, and extra-vehicular activity.

The final phase is the 'Special Skills' training block, which includes modules on robotics, rendezvous and docking, and Russian language. With this phase complete, ESA astronauts begin more specific training – ISS Advanced Training – geared to teaching them how to operate the many systems and subsystems on board the ISS and on board the Soyuz. This phase takes about another year and includes visits to the various training centers in Houston, Star City in Russia, Tsukuba in Japan, and Montreal in Canada. Once all this training is completed, astronauts are finally eligible for a mission, and once they get their mission assignment they begin another 18 months of training, called Increment-Specific Training (IST). The IST phase teaches the astronauts what they need to know for their mission, and for the lucky few who have been assigned spacewalking duties, time is spent in the Neutral Buoyancy Laboratory (NBL – Figure 5.2).

So that's how government astronauts are trained. Now let's turn our attention back to Mars One.

"Greetings, candidates. The testing that begins today will determine which of you have the necessary skills to thrive on Mars. Think of how satisfying it will feel to stand in the Valles Marineris and thunder-chug a Blue Bronco energy drink. I'm sorry, Blue Bronco? Wait, thunder-chug? Blue Bronco is just one of this mission's many corporate partners. Because this is a privately funded entrepreneurial mission, we've teamed up with some of America's most exciting brands. We're talking Mega-Charge Batteries, Fantasy-Lunatics.com, Trudge-Rite Work Boots, Draft-Pigs. And who here likes Fig Glutens?"

The Simpsons, Season 27 Episode 16, The Marge-ian Chronicles, *Original Air Date March 13, 2016: in this episode Lisa signs up for a one-way mission to Mars – an episode inspired by a very well known reality TV concept!*

Sorry, but I couldn't resist; I'm a *Simpsons* fan! Now where were we? Oh yes, technical training. Let's begin by re-emphasizing this statement from the Mars One archives:

"Technical training includes the training of two crewmembers to be proficient in the use and repair of all equipment to the extent that they can identify and solve technical problems."

Two? I'm sorry, but two will not cut it. Not by a long, *long* shot. That's because the spacecraft that will transport the Mars One contestants to their new home will in effect be a fast-moving home improvement project that will require maintenance and repair around the clock. How do we know this? Because the ISS has been zipping around up

Figure 5.3   Housekeeping on board the International Space Station. Credit: NASA

on a one-way trip to Mars, a failure of the CDRA, with no options for repair, represents more than an operational problem doesn't it? That's a death sentence, pure and simple. Now perhaps the Mars One crew will take along a 3-D printer and be able to print new parts, but that technology is still a way off. And if they do load a 3-D printer on board, they should load plenty of printing material because, as I mentioned before, the spacecraft will be an ongoing DIY project. Just like the ISS. A couple of months after the ASV issue, the crew had to deal with a failure of one of the cooling systems. The cooling system comprises a pump, a radiator and some pipes filled with ammonia. The function of this system is to take heat from the interior of the ISS to the exterior, where it is dumped into space. When one loop fails, the back-up kicks in, but as a precaution, many systems have to be shut down. Sometimes the problem is a flow valve malfunction. On other occasions, a preset threshold temperature is reached which triggers a shutdown. Fortunately, like the ASV, the ISS has a redundant cooling system, otherwise the crew would be bailing back to Earth on board the Soyuz on a regular basis. That won't be an option for the Mars One quartet. So let's hope the Mars One astronaut training includes a lot of on-systems technical troubleshooting (see sidebar). A lot.

The 'Big 14'.

The ISS has so many problems that NASA requires each crewmember to be trained to repair the most common ones – the 'Big 14'. The following list of these critical maintenance tasks has been sourced from the Increment Definition and Requirements Document. And a note about the use of the word 'critical' to those in the Mars One organization who may not be familiar with NASA parlance: 'critical' means zero tolerant, which is NASA-speak for a failure of that particular system being zero tolerant for survival. Got that? Good.

1. Maintain ISS Primary Electrical Power System Survivability

    a. External (EXT) Multiplexer/Demultiplexer (MDM) Remove and Replace (R&R)
    b. Battery Charge/Discharge Unit (BCDU) Backout
    c. Main Bus Switching Unit (MBSU) R&R
    d. Sequential Shunt Unit (SSU) R&R
    e. Direct Current Switching Unit (DCSU) R&R
    f. DC to DC Converter Units (DDCUs) R&R
    g. Solar Array Wing (SAW) Manual Positioning
    h. Pump Flow Control Subassembly (PFCS) R&R
    i. Photovoltaic Controller Unit (PVCU) MDM R&R
    j. External Remote Power Control Modules (RPCMs) R&R

2. Maintain ISS Thermal Control System Survivability

    a. Interface Heat Exchanger (IFHX) R&R
    b. External Thermal Control System (ETCS) Pump Module (PM) R&R
    c. Flex Hose Rotary Coupler (FHRC) R&R
    d. Ammonia (NH3) Leak Isolation and Recovery

How many 'Big 14' items will the Mars One crew need to be trained on? Probably a hell of a lot more than 14. All we can do is hope that training is first-rate and that the crew is capable of fixing what will be a seemingly endless list of problems. If not, then the reality show that is Mars One will be broadcasting Apollo 13 levels of drama. Great for the audiences. Not so great for the contestants. Think I'm exaggerating? Then consider the following. More than 200 astronauts have lived on board the ISS, and all that time on orbit has taught engineers on the ground that life support systems break down. All the time! You might think that after spending the best part of $120 billion on the orbiting outpost we could expect a higher level of reliability, but that's spaceflight life support for you. For BEO human exploration, this hardware must perform reliably for a long, long time, under much harsher conditions than in Low Earth Orbit (LEO). At least if a fix can't be found on board the ISS, the crew can simply bundle into one of the aging Soyuz capsules and take a ride back to Earth. That won't be the case for the Mars One contestants, so technical training and troubleshooting skills will be critical. You see, in space, when it comes to trouble, it never goes away. Not permanently. Trouble is always in the next module doing press-ups. Now, what's next on the Mars One astronaut training agenda? Oh, yes. Medical training.

SPACE DOCTORS

In the less than informative publication that is *Mars One: Humanity's Next Great Adventure*, there is a chapter titled 'Medical Skills for an Interplanetary Trip'. In the chapter, the author, one of the world's leading experts when it comes to space life sciences, outlines some of the medical skill-sets the contestants will probably find helpful during their very brief mission. Surgery is mentioned, as is the challenge of dealing with the deleterious effects of living in a reduced gravity environment. True, the contestants will definitely need to be able to handle those problems, but what else will they have to deal with? Let's take a look shall we?

One of the myriad in-flight emergencies our contestants will most likely face will be radiation sickness. Let's face it, even NASA and ESA haven't solved the problem of protecting crews from radiation and the Dollar Store enterprise that is Mars One certainly won't be able to afford to protect their contestants. Perhaps it will be for lack of shielding, or perhaps a solar flare will erupt, but without sufficient radiation protection, radiation sickness is more or less guaranteed. Let's go with the solar flare scenario. Perhaps three of the crew manage to shield themselves by cramming themselves in between the water containers, but the remaining crewmember takes a big hit. At this stage, let's think of some suitable reality TV names for this imaginary crew. Hmm. Let's see. Kendra can be our engineer, Selena can be our pilot, Josh can be our commander and Simon can be our scientist. We'll pretend it was Kendra, our plucky engineer, who took the radiation hit. What do Selena, Josh and Simon do? Well, to begin with they will need to determine the absorbed radiation dose, because it is this information that determines how severe the sickness will be and which treatments they will have to administer. In between Kendra's vomiting and howls of excruciating pain, the symptom-free contestants will need to keep their cool and figure out how much radiation their crewmate absorbed. To do this, they may consult the information presented in Table 5.1.

But knowing the symptoms will only be part of the story when it comes to saving Kendra. The next step will be diagnosing what Acute Radiation Syndrome (ARS) she has.

Table 5.1 Signs and symptoms of radiation sickness[1]

	Mild Exposure (1-2Gy)	Moderate Exposure (2-6Gy)	Severe Exposure (6-9Gy)	Very Severe Exposure (10Gy or higher)
Nausea/vomiting	Within 6 hrs	Within 2 hrs	Within 1 hr	Within 10 mins
Diarrhea	N/A	Within 8 hrs	Within 3 hrs	Within 1 hr
Headache	N/A	Within 24 hrs	Within 4 hrs	Within 2 hrs
Fever	N/A	Within 3 hrs	Within 1 hr	Within 1 hr
Dizziness	N/A	N/A	Within 1 wk	Immediate
Weakness	Within 4 wks	Within 1 – 4 wks	Within 1wk	Immediate

[1]Based on Radiation exposure and contamination, Merck Manual, Professional Edition

The three classic syndromes of ARS are bone marrow, gastrointestinal (GI) and cardiovascular/central nervous system syndrome (a fourth is cutaneous radiation syndrome). In each case, the chances of survival are slim. A patient with bone marrow syndrome will mostly likely die from the destruction of bone marrow and consequent infection and hemorrhage, while a patient suffering from GI syndrome will suffer irreparable changes to the GI system and consequent catastrophic electrolyte imbalance. A patient with either of these syndromes can expect to live no more than two weeks; if they are attended to by nurses around the clock in a full-spectrum terrestrial Intensive Care Unit (ICU). There won't be one of those on a Mars-bound spacecraft, though. If she – and the crew – is lucky, Kendra will be a victim of cardiovascular/CNS syndrome. She will be lucky because this syndrome kills within three days, tops, and Selena, Simon and Josh will thank their lucky stars because they won't have to expend any more life support consumables attending to her. But whichever syndrome Kendra is diagnosed with, the crew will have their work cut out for them, because treating a patient with ARS is unpleasant, even for highly trained medical personnel. First, there is the prodromal stage, in which Kendra will spend most of her time projectile vomiting and being sick with diarrhea. Imagine cleaning all that up in zero-G! And this stage could last for days, requiring the crew to attend to their ailing crewmate 24/7. High drama on television back home though. In the latent stage Kendra will appear healthy, but this is a false dawn.

After a few days or weeks, Kendra will enter the manifest illness stage, in which the symptoms will depend on the syndrome. This stage could last for hours or it could last for months, but sure as eggs are eggs, Kendra is going to die. If she had been on Earth she would have had a chance of recovery, but in deep space the kindest option will be euthanasia, because the medical supplies of an interplanetary spaceship will be quickly overwhelmed. Let's take it step by step and begin with triage. First the crew will need to secure the ABCs – airway, breathing and circulation – while monitoring Kendra's cardiovascular parameters and urine output. Next, they will need to treat any major trauma, such as burns, and also take blood samples to determine the lymphocyte count. Treatment will be intensive, and in short order the spacecraft's medical support system will be overwhelmed, leaving the crew (and the audience back home on Earth) to watch in pity as Kendra's skin peels, her white blood cell count plummets and she spews out her shredded intestinal lining. Quite the clean-up operation! But the crew will keep fighting – at least we hope they will – because that is what reality television show contestants do. Back home on Earth, the blogs will be alive with messages of 'I told you so' and the producers will have a ratings bonanza on their hands. Meanwhile, millions of miles from home, Kendra's crewmates will settle into a routine of taking photos of Kendra's skin to record radiation damage, initiating viral prophylaxis and consulting with radiation and hematology at mission control to be advised on dosimetry, prognosis and treatment options. That treatment will have the goals of preventing further radioactive contamination, treating life-threatening injuries and managing the pain. This is where it will get even more messy, so let's hope Mars One has the foresight to train the lucky four in decontamination procedures, because the first order of business will be to remove external contamination by washing Kendra's skin with soap and water, all the while fretting over the hemorrhaging of life support consumables for what will ultimately be a futile gesture of goodwill. At this stage, the thought of flushing Kendra's body through the airlock will have crossed someone's mind. But like good little reality television soldiers, the crew will fight on. Next on the 'to do' list will be to treat Kendra for damaged bone marrow. Hopefully the crew will have brought some granulocyte colony stimulating factor along (this could be medication such as Neupogen, Leukine or Neulasta), because it is this protein that promotes the growth of white blood cells, which may offset the effect of ARS on Kendra's bone marrow. Transfusions? Sorry honey-bun, but we're in deep space. Since Kendra has suffered internal damage, the crew will need to administer potassium iodide and perhaps Prussian Blue, which is type of dye that binds to radioactive elements. This course of treatment could go on for days. Or weeks.

All the while, Kendra will suffer headaches, fever, nausea, bacterial infections... In short, her life will be miserable and since she absorbed a dose of more than 10Gy, the best the crew can realistically do is perform end-of-life care. This will be a first for manned spaceflight, so chalk one up for Mars One on that count. But what will the audience at home think of this? If ever there was a case for euthanasia, it would be now. Will Mars One have prepared the crew for burial in space? How would this be done? Over to you Mars One, any suggestions?

DEATH IN THE VOID

Like so much that is Mars One, nobody in the organization has a great plan for what to do when one of the contestants dies. Having said that, even NASA doesn't have a protocol for what happens when an astronaut dies up there, although astronauts do practice for worst-case scenarios – Mars One take note! One such exercise is called the 'death sim'. Here's how this (will) might play out during Mars One. The Mars One mission control will receive the message: 'Kendra is dead'. After the initial shock, someone might ask what should be done about the corpse, because there will be no body bags on the spacecraft. Do the three surviving contestants bundle Kendra's body into a spacesuit and stuff it in a locker? Probably not, because the smell would be rather... funky, after a while. Send it back in a resupply ship then? No, can't do that. It's a Mars mission remember. Flush it through the airlock then? Now you're talking! And think of the ratings! The first burial in space – think Spock's funeral in 'Star Trek' – and an invaluable training opportunity to boot. After all, the other contestants won't have long to live anyway! So, as the contestants busy themselves with opening the airlock, the Mars One reality show can call Kendra's mum and dad and tell them that their daughter's body is about to be flushed into the void. Another space first! Problem solved? Not quite. You see, there is the business of space law that says you can't dump your litter in space, and that includes the corpses of dead reality show contestants, even if said contestants have expressed a wish to have their body committed to the void. You may be wondering why dumping litter in the void that is space would be banned, but there are reasons. One reason is that garbage can collide with other spacecraft (unlikely on a Mars transit) and another reason is that a body could eventually float over an alien planet and colonize it with the bacteria still living in the corpse (even more unlikely). Remote chance, but still possible according to those that make up these rules. So Mars One needs a Plan B. Keeping the body on board wouldn't be good for the crew's health or morale and adding a locker to store corpses would be cost-prohibitive, so what other options are there?

"The method behind ecological burial is called 'Promession' and it is crystal-clear, easy to grasp and accept. It is based on a new combination of tried-and-tested techniques that prepare the corpse for a natural process of decomposition. The procedure is justifiable in terms of ethical, moral, environmental and technical considerations, and does not subject the body to violent or destructive handling."

Excerpt from www.promessa.se.
Promessa was founded by Susanne Wiigh-Mäsak in 2001

Promession is a technique developed by the ecologically-friendly burial company Promessa. Very simply, using this technique would require the Mars One contestants to zip Kendra's body into an airtight sleeping bag, flush it out of the airlock and attach it to the vehicle. Once frozen, Kendra's body would be vibrated until her corpse shattered. Result: about 20 kilograms of human dust, which the Mars One contestants would put into a ZipLock bag before attaching the bag to the outside of the vehicle. Then, after landing, Kendra's remains could be used as fertilizer and compost material for growing crops. What an elegant way to die! Taboo? Unethical? Not at all. This is a pioneering mission remember. Yes, there is a chance that fertilizing Martian soil using body-based fertilizer may cause environmental contamination, so I suggest Mars One contact NASA's chief bioethicist, Dr Paul Wolpe, for guidance on the subject. Look him up at the Emory Center for Ethics. Got that?

"We don't know much about what kinds of highs and lows we'll see in people over long periods of time, under extreme circumstances."

Steve Kozlowski, PhD psychologist at Michigan State University

"I think these will be bigger challenges than technology challenges."

Jason Kring, Embry-Riddle Aeronautical University, Florida

COPING SKILLS

So, we've covered technical training, medical training and bioethics. What's next? Well, Phase 2 mentions coping skills and Phase 3 mentions simulations. First, coping skills. For years, we have been bombarded by the dire predictions of psychologists agonizing over

Figure 5.4 Coping Skills. History is full of stories of explorers who survived much harsher conditions than anything that will be faced during a trip to Mars. Pictured is one of the bravest crews who ever set forth on an expedition: Shackleton's Imperial Trans-Antarctica crew who spent a long time isolated on Elephant Island. A trip to Mars will be a picnic compared to what these guys went through. A picnic. Public domain

the challenges of isolation faced by a crew en route to Mars. These 'Mars crews will go crazy' mythologists insist that the psychology of leaving Earth behind will result in astronauts losing their minds, and that lots and lots of research is needed to figure out how to deal with the challenges. They insist psychology is perhaps the toughest challenge of all. Well, if you believe that then you probably believe humans shared the Earth with dinosaurs, because one thing we happen to know an awful lot about is isolation (Figure 5.4). You see, 100 years or so ago there was what was known as the 'Heroic Age of Antarctic Exploration', a term coined by British explorer Duncan Carse in March 1956 when describing the exploits of Sir Ernest Shackleton.

This Heroic Age began in 1897 with an expedition funded by the Belgian Geographical Society, and ended either with Shackleton's *Endurance* expedition in 1917, or when the great man died in 1922, depending on which historical text you happen to be reading. Why is the Heroic Age so important? Because back in those days, explorers were explorers and ventured into the unknown for years at a time, all without the benefit of GPS, iPhones, social media, e-mail, reality television shows or daily updates on the *Kardashians*. Years! And the time they spent away from their family and friends was spent in the most austere environments imaginable. Blizzards, winds in excess of 200 kilometers per hour, and temperatures that plunged below minus 70°C. A trip to Mars? Compared to what the explorers in the Heroic Age went through, a trip to the Red Planet will be a luxury cruise, even if it is a one-way deal. But despite the Heroic Age having been the Space Age of its day, and despite dozens and *dozens* of books having been published on the various adventures of Amundsen (Figure 5.5), Nansen, Mawson and Shackleton, it seems that psychologists nowadays are oblivious to this wealth of knowledge about how humans dealt with extreme isolation against the odds. And the reason we have this wealth of knowledge about the psychology of survival more than 100 years ago is because these explorers kept meticulous diaries, many of which can be found on the bookshelves of libraries around the world. For the benefit of those unfamiliar with the expeditions of the Heroic Age, what follows is a snapshot of a group of explorers who suffered extreme isolation.

Figure 5.5 Roald Amundsen. One of the greatest ever explorers. Public domain

"Polar exploration is at once the cleanest and most isolated way of having a bad time which has been devised."

Apsley Cherry-Garrard, The Worst Journey in the World.

In 1893, Fridtjof Nansen sailed to the Arctic in the *Fram* (Figure 5.6), a purpose-built, round-hulled ship that was the Space Shuttle of its day. The *Fram* was designed to drift north through the sea ice in the footsteps of the *Jeannette*, which had foundered northeast of the New Siberian Islands and was found on the southwest coast of Greenland after having drifted across the Polar Sea. Nansen reckoned the Polar current's warm water was the reason for the movement of the ice. But, after more than one year in the ice, with the *Fram* going nowhere, plans had to be changed. When the *Fram* reached 84° 4´ North, Nansen, accompanied by Hjalmar Johansen, headed north on foot. It was a bold move, as it meant leaving the *Fram* not to return, and a journey on foot over drifting ice to the nearest known land, 800 kilometers south of the point where they started.[1]

Figure 5.6 Fridtjof Nansen's *Fram*. Public domain

Nansen and Johansen started their journey on March 14, 1895, with three sledges, two kayaks and 28 dogs. On April 8, 1895, they reached 86° 14´ N, the highest latitude ever reached at that time. The men then turned around and started back, but they didn't find the

---

[1] All appear in: *Survival and Sacrifice in Mars Exploration: What We Know from Polar Expeditions*, published by Springer Praxis Books in March 2015. ISBN-13: 978-3319124476.

land they expected. On July 24, 1895, after using their kayaks to cross stretches of open water, they came across a series of islands, where they built a hut of moss, stones, and walrus hides (Figure 5.7). Here, they spent nine mostly dark months, sleeping for up to 20 hours out of every 24 while waiting for the daylight of spring. They survived on walrus blubber and polar bear meat. In May 1896, Nansen and Johansen made tracks for Spitsbergen. After traveling for a month, not knowing where they were, they were delivered from their endeavors through a chance meeting with Frederick George Jackson, who was leading the British Jackson-Harmsworth Expedition which was wintering on the island. Jackson informed them that they were on Franz Josef Land. Finally, Nansen and Johansen made it back to Vardø in the north of Norway.[1]

Figure 5.7 The hut where Nansen and Johansen spent nine months isolated in some of the most extreme conditions. A little tougher than the nine-day isolation test planned for the Mars One contestants! Public domain

It is hard to imagine a more primitive confinement than Nansen and Johannsen's winter of 1885 to 1886, but the annals of polar exploration are rich with resilient and resourceful individuals. Take the gripping tale of four Pomori hunters who, in 1743, found themselves marooned on Edgeøya Island of the Svalbard Archipelago (Figure 5.8). For six years, this group of hardy individuals survived everything the Arctic could throw at them: storms, chilling cold, extraordinary deprivation, confinement, and polar bears. Their story started when the four sailed as part of a group of 14 hunters from the village of Mezen on the White Sea coast. They planned to hunt walrus in the Svalbard Archipelago. After eight days with favorable weather, they were blown off course towards Edgeøya Island, a place where ships rarely ventured. Before long their vessel was icebound. The situation deteriorated over the next few days as it seemed likely their vessel would be crushed. It was decided that a four-man party would go to the island to investigate what shelter there was, since it was known sailors had spent the winter there several years previously in a hut. The four knew that they wouldn't be gone long and that the hunting would be excellent, so they carried only the barest essentials. On reaching the island they found the hut, where they spent the night while a storm blew outside. The next day, they made their way back to their ship to share the news with their fellow hunters. But on reaching the shore, they discovered that part of the ice pack had gone, and with it the ship, presumably carried away by the storm the previous night. The four returned to the hut and pondered the likelihood that they were now trapped on the island – permanently. The kept a watch for their ship, but after a few days they came to the conclusion that it had foundered (the ship never returned to port, so the assumption was probably correct)[1].

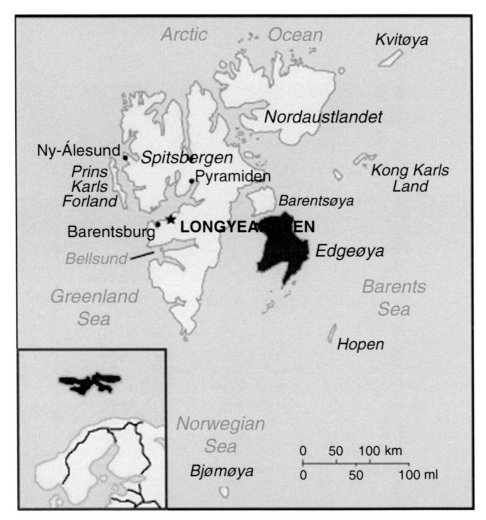

Figure 5.8   Edgeøya. Public domain

The four faced a bleak existence, stuck on an island in the middle of a polar bear breeding ground and with all their ammunition expended. Worse, the island was devoid of trees or shrubbery, which meant they had nothing to burn and couldn't cook. But they resolved to try to survive. After all, what else could they do? They scoured the island and found some driftwood, as well as a plank with a long iron hook attached and some embedded nails. It wasn't much, but for these guys it was a lifeline. Using a primitive forge, they fashioned their newly-found hardware into a sharp point, which they attached to a driftwood pole. The Pomori now possessed a weapon and set out to hunt polar bears. After killing and eating their first bear, they cut the skin to be used as clothing and used the animal's tendons to create a string for a bow. The nails were used to manufacture rudimentary arrows. With their bow and arrows, the Pomori killed more than 250 reindeer, along with an assortment of foxes. As winter closed in, fuel economy became vital, but they couldn't allow the fire they had set to go out. Fortunately, there was no limit to the Pomori's ingenuity. They gathered some of the slimy loam they had found during their reconnaissance of the island and fashioned a lamp. Reindeer fat was placed into the lamp and that became their source of warmth during the long winter. Another challenge was food. The only vegetation on the island was moss and lichen, so the men subsisted on reindeer, fox and bear. Water was drawn from springs and made by melting ice. To prevent scurvy, the men drank reindeer blood and ate the little grass that grew on the island. Their psychological condition? Who knows? How did they deal with the long years of confinement and paralyzing, mind-numbing monotony? How did they cope with the chilling cold and the appalling condition of the smoke-filled hut? Who knows, but the experience of these castaways serves as an important reference point for the Mars One contestants who will be confined in a Mars-bound spaceship.

The Pomori were eventually rescued when, on August 15, 1749, the hunters caught sight of a Russian trading vessel on its way to Novoya Zemlya. The ship had been blown off course and had inadvertently found itself near Edgeøya. Six weeks later, the men were finally returned home. What these hunters achieved serves as a lesson to all explorers about faith, perseverance, ingenuity and resourcefulness. Which is why many question the utility of mission simulations such as the Mars500 boondoggle, which placed six humans in a tin can for 520 days to simulate a space mission. Why? Did these researchers bother to read Nansen's *Farthest North*? Were they aware of the Pomori's extraordinary ability to survive? Obviously not. There are dozens and *dozens* of accounts of humans surviving in isolation in austere environments, in much more primitive conditions than a nice air-conditioned spaceship, so there will probably be no need for Mars One to worry about how the crew will fare in isolation. What they *should* be concerned about is how their crew functions as a group when faced with a challenging environment. For that we need an analog, which happens to be one of the training increments mentioned in the Mars One blurb. But not just any analog. This one has to be realistic, unlike the Mars500 boondoggle or the groups living in Hawaii. Hawaii! To simulate a space mission? Seriously? No, the analog mission I'm thinking of will be a real simulation. We'll call it the Nansen Sim (Figure 5.9).

"From its inception, space exploration has pushed the boundaries and risked the lives and health of astronauts. Determining where those boundaries lie and when to push the limits is complex. NASA will continue to face decisions as technologies improve and longer and farther spaceflights become feasible. Our report builds upon NASA's work and compiles the ethics, principles and decision-making framework that should be an integral part of discussions and decisions regarding health standards for long duration and exploration spaceflight."

Jeffrey Kahn, Chair of Committee, Health Standards for
Long Duration and Exploration Spaceflight.

Figure 5.9   Orion. Credit: NASA

Mr. Kahn is right. We need to push the limits, which is the purpose of the Nansen Sim. Until a faster propulsion system is operational, a Mars One mission will take about seven months plus a little time on the surface: let's be optimistic and give them a life expectancy of 10 weeks after landing. So the Nansen Sim will begin with a seven-month period orbiting the Moon, followed by a surface stay of 10 weeks. Why not orbit the Earth you may ask? Well, we want to make sure our Mars One contestants are exposed to as much deep space radiation as possible. Just like they will be on the actual mission. Remember, we're striking out and being bold, just like those explorers in the Heroic Age. Something that Shackleton might have appreciated. And Nansen. Here's how the mission might play out. After being sterilized and having their appendices and gallbladders removed, our four contestants will launch into Low Earth Orbit (LEO) and then head for the Moon, where they will spend seven months in orbit. During their time in lunar orbit they will be denied communication with friends and family, to see how this group fares when truly isolated. Then, after the seven months are up they will land on the Moon and practice the tunneling techniques that will be required to keep them safe from the deadly radiation. When the 10 weeks are up they will return to Earth, where they will be studied.

In addition to being a survival mission and a test of the crew's resistance to the effects of isolation, the mission will also include several scientific objectives. One field of study will be behavioral health, which will examine such factors as sleep and neurocognitive performance (there won't be any studies examining the effects of isolation though, because we know how humans perform in isolated environments for prolonged periods of time). Another field of study will examine the extent of visual impairment on the crew. To date, NASA only has data on crewmembers who have spent six months on orbit. How much worse will the visual impairment problem be after 30 months? Thanks to the Mars One test crew we will find out: don't forget that the crew will be exposed to deep space radiation and therefore there is a good chance some or all of the crew will develop cataracts. How will they deal with that problem? No abort or rescue capability on this mission remember – just like there won't be any rescue on a real mission to Mars. A third field of study will be metabolic investigations aimed at defining the extent of oxidative stress during what will be a very long mission, while the fourth series of investigations will examine physical performance. This will be particularly interesting because we know some astronauts lose as much as one percent of their bone mass per month, so there is a good chance some of the crew could lose up to a third of their bone mass by the time they return to Earth. Will they be able to conduct all their mission tasks on the surface of the Moon without risk of fracture? We won't know until the crew embarks on this mission. The fifth field of study will be human factors, examining how this pioneering group interact with their environment en route to the Moon and while they work on the lunar surface. And finally, the sixth area of investigation: rehabilitation post-flight. Can this crew be rehabilitated? Will they even be able to walk again? Good to know the answers to these questions if you plan on having a reality show wouldn't you think? As for the psychological assessment? Well, there's no need for post-mission psychological studies because this crew will have spent the duration of their mission in the proverbial lap of luxury compared to the travails of Shackleton, Cherry-Garrard and company.

> "The horror of the nineteen days it took us to travel from Cape Evans to Cape Crozier would have to be re-experienced to be appreciated; and anyone would be a fool who went again: it is not possible to describe it… It was the darkness that did it. I don't believe minus seventy temperatures would be bad in daylight, not comparatively bad, when you could see where you were going, where you were stepping, where the sledge straps were, the cooker, the primus, the food; could see your footsteps lately trodden deep into the soft snow that you might find your way back to the rest of your load; could see the lashings of the food bags; could read a compass without striking three or four different boxes to find one dry match; could read your watch to see if the blissful moment of getting out of your bag was come without groping in the snow all about; when it would not take you five minutes to lash up the door of the tent, and five hours to get started in the morning."
>
> *Excerpt from* Worst Journey in the World. *Apsley Cherry-Garrard*

TEDDY BEARS AND BOONDOGGLING

So please: no more wailing about how isolated these Mars One contestants will be. No more griping about how real time interaction with people back home will be impossible, and no more whining about difficult it will be for these contestants to be faced with the prospect of only interacting with each other for the rest of their lives. Look, in the best case scenario, the chances are high that they will never reach the surface of Mars, so 'the rest of their lives' won't amount to a tin of beans. And no more of this nonsense about how isolation will cause mental illness, depression, insomnia, and emotional instability. Have 21st century humans truly become so soft that people really believe a few months cooped up in cozy spaceship will cause nervous breakdowns? Apparently the psychologists do. Shackleton, Nansen and company must be spinning in their graves. Look, all these so-called experts should stop making sensationalist comments about crewmembers becoming homesick and isolated and start reading some of the classics of exploration. *'Endurance'* by Alfred Lansing, *'Fatal Passage'* by K. McGoogan, *'The Worst Journey in the World'* by Apsley Cherry-Garrard – take your pick. Why? Because by reading these books they will discover that humans have an extraordinarily high threshold for isolation and confinement. At least, 100 years ago they did. The problem these days is that people are soft (see sidebar). They have their iPhones and streaming video. They have their on-demand movies and e-mail. They have their… well, you get the idea. Mention 'social isolation' and today's merchants-of-doom psychologists throw their collective teddy bears in the corner and complain about the terrible health burdens that will result. Oh please!

Figure 5.10   Landing on Mars, Mars500 style. Credit: ESA

Boondoggled #1

To boondoggle is *to perform work of little or no value, merely to keep or look busy.* Here's an example. In 2005 the US government authorized $452 million to build two bridges in Alaska. One of these bridges, which became known as the Bridge to Nowhere, would have connected Ketchikan to Gravina Island, home to just a few dozen people. Crazy, right? That's a boondoggle for you. Here's another. In July 2009, the State Scientific Center of the Russian Federation conducted a 105-day pilot confinement study by locking six crewmembers in a tin can (see Figure 5.10). This study was a prelude to a 520-day study that was supposed to simulate a mission to Mars. Completed in November 2011, six multinational crewmembers spent 520

(continued)

(continued)

consecutive days of confinement in a 550 m³ pressurized facility. Facility modules were equipped with life support systems and an artificial atmospheric environment at normal barometric pressure, and activities simulated the work routine on board the ISS. The crew lived on a five-day work cycle, with two days off, except for emergency simulations. Dozens of experiments were performed in the disciplines of physiology, biochemistry, immunology, biology, microbiology, operations and technology, and of course psychology. I wonder what the Pomori would have made of these crewmembers living in the lap of luxury? The Mars500 crew completed depression inventories to see how suicidal or irritable they were, and also completed conflict questionnaires to determine when crewmembers argued the most. Once the mission was over the scientists had reams of data, but did they actually learn anything? Well, the scientists discovered that crewmembers exhibited depressive symptoms and some psychological distress, but nothing that hasn't been observed *hundreds* of times before in polar explorers. Sleep-wake data revealed insomnia in some crewmembers, which researchers suggested could be detrimental during critical periods of the mission, such as docking maneuvers or responding to emergencies. Perhaps, but polar explorers were subject to extended periods of insomnia, compounded by the most horrendous conditions imaginable, and were still able to deal with critical tasks. Take Worsley's epic feat of navigation in Shackleton's *Endurance* expedition that was accomplished in icy storms and 30-meter high waves. Enough said.

Boondoggled #2

Boondoggling doom merchants and what they have said about traveling to Mars:

"When a bad day happens, it isn't so easy to address in space. It's inherently difficult, psychologically, to make sure astronauts are able to handle this."

Clinical psychologist and senior scientist at the SETI Institute.

"The absence of a visible home planet may result in increased feelings of isolation, homesickness, dysphoria, or even suicidal or psychotic thinking. Mars might entrance astronauts, but it isn't home. Imagine an astronaut wants to phone home and say 'Houston, I'm having a problem'. Well, they'll be waiting an awfully long time before they can get any advice."

Professor of psychiatry at the University of California, warning about the terrible things that will happen on the way to Mars.

"We don't know much about what kinds of highs and lows we'll see in people over long periods of time, under extreme circumstances. This should give us some understanding of what we can expect."

Researcher, trying to justify simulating a Mars mission in a domed habitat on Hawaii's Mauna Loa volcano. Hawaii? Really!

The boondogglers, sorry, researchers, attempted to justify their research by stating the importance of identifying psychological markers that predispose long-duration crewmembers to psychosocial reactions during the confinement required for exploration missions. They went on to say that such predictors and biomarkers are needed to select and train crews and that Mars missions will require the 'right stuff' for prolonged confinement and isolation. Well, they're right about that, but a trip to the local library could have told the researchers everything they needed to know about man's capacity to survive in isolation and confinement. No need to lock a crew up in a tin can for 17 months. Ultimately, the Mars500 analog was extremely limited, not only by its absence of zero gravity, but because of the very generous comfort blanket available to the crew, who could have left the module at any time. The Mars One contestants will have no such option. So, based on the results of Mars500, were researchers able to answer the question "Is man able to endure the confinement of a trip to Mars?" No. Not by a long shot. But, based on the experiences of the Pomori, Shackleton, Nansen and Co., man most certainly is.

Here's another boondoggle for you. A few years ago, NASA handed out $1.3 million to psychologists to develop a psychosocial sensing badge that astronauts might wear during a Mars mission. I swear I'm not making this up. This badge was designed to monitor which crewmember approached another, how long their conversation was, whether vocal patterns were friendly or acrimonious, and who ended the discussion. Sensors would then provide feedback to the crewmembers about their behavior. Perhaps a sensor would detect aggressive behavior and recommend that the astronauts make peace. Computerized therapists in other words. Are we treating astronauts like babies now? Give me a break.

For many psychologists, traveling to Mars is all doom and gloom, hence the need for more boondoggle studies. These studies are completely unnecessary because crewmembers are very unlikely to get bored on a trip to Mars for one very good reason: *salutogenesis*. It is a term coined by Aaron Antonovsky, a medical sociology professor who wanted to come up with a word which conveyed the idea that, under certain conditions, stress could be beneficial and health-promoting, and not pathogenic or destructive to health. As you can imagine, polar explorers experienced all sorts of negative effects as they struggled to cope with isolation, deprivation and extreme conditions. But on the flip side, the elation of having coped with so much successfully brought positive benefits. So explorers tended to enjoy the experience and enjoyed positive reactions to the challenges of the environment. Not only that, but this unique group of individuals actually thrived on the feeling of having successfully overcome these challenges. In their diaries (if any of these doom and gloom psychologists are reading this, I suggest they read some of these accounts), they routinely refer to the beauty and grandeur of the land, ice, and sea, the camaraderie and mutual support of the team, and the thrill of facing and overcoming the challenges of the environment. Which is probably why so many signed up for repeat expeditions.

But our pessimistic psychologists fixate obsessively on the deleterious effects of long duration missions, paying scant attention to the beneficial effects of such an endeavor. Which is a shame, because polar exploration has shown that individuals who adapt positively to an inhospitable or extreme environment can derive benefit from their experiences (Table 5.2). And this positive effect may include an initial improvement in mental health as a crewmember adapts to the environment.

Table 5.2 Salutogenic after-effects of polar expeditions

Sense of personal achievement
Striving toward important goals
Courage, resoluteness, indomitability
Excitement, curiosity
Increased self-esteem
Hardiness, resiliency, coping
Improved health
Group solidarity, cohesiveness, shared values
Increased individuality, reduced conformity
Ability to set and achieve higher goals, and changes in thinking

Despite most researchers choosing to ignore the salutogenic effects of spaceflight, these effects have been observed during missions. Astronauts report positively about the friendship and cohesiveness among the crew, satisfaction in jobs well done, pride in having been chosen to fly their mission, and an appreciation of the beauty of Earth from space. In fact, the current trend in memoirs written by spacefarers is to refer to positive emotions three times as often as to negative ones; a good recent example being Chris Hadfield's *An Astronaut's Guide to Life on Earth*. Astronauts' autobiographical accounts routinely mention trust in others, autonomy, initiative, industry, strong personal identity, and a conviction that their life makes sense and is worthwhile. These astronauts were confident about their emotional stability and coping abilities, and viewed themselves as active agents in dealing with problems – just like Shackleton and his crew, or Nansen and Johansen. These autobiographical reports point to some inescapable conclusions. First, space agencies select resilient people who are good at solving problems and getting along with others. Second, for most astronauts, spaceflight is their peak life experience. Third, examining post-flight changes, astronauts consider themselves to be changed for the better. These findings in no way detract from the importance of anticipating problems and preparing countermeasures for the unique challenges of a Mars mission, but equally, they underline the importance of also considering the possibly unique benefits of this great adventure, to the astronauts themselves and to humankind.

6

Medical mission-killers

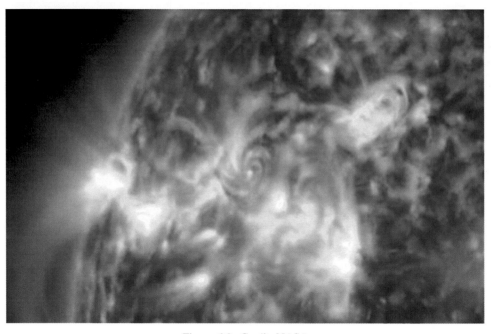

Figure 6.0   Credit: NASA

Extrapolative science fiction. It's entertaining stuff; whether you happen to be watching *Blade Runner* and thinking there might be Replicants running around one day, or appreciating the drama of *The Martian* and pondering the possibility that science fiction might just become science fact. But one of the common failures of extrapolative sci-fi is the assumption that just because something might be feasible, it will happen. Which is exactly the case with a manned mission to Mars, whether this be a government-sponsored enterprise or a private one-way venture. A manned mission to Mars is a perennial favorite topic

© Springer International Publishing Switzerland 2017

E. Seedhouse, *Mars One*, Springer Praxis Books, DOI 10.1007/978-3-319-44497-0_6

that somehow keeps on coming back from the dead: we've been 30 years away from such a venture for the past 50 years and we're still not any closer... unless Elon Musk has something up his SpaceX sleeve. I'm not saying it won't happen; I'm just saying it may not happen for a long, *long* time. And this chapter explains why.

Figure 6.1 The Next Giant Leap. Credit: NASA

REALITY BITES

One reason Mars is kept in the spotlight is because the Red Planet evangelists insist that a manned mission is just around the corner. Another reason public enthusiasm is kept alive is thanks to the occasional landing of a rover such as *Curiosity*. Such events inevitably prompt the spaceflight community to gather and discuss plans for a manned mission to Mars. Conferences such as the Humans to Mars (H2M), for example. The H2M get-together, which was hosted by the advocacy group Explore Mars, announced it was calling for a manned mission in the 2030s. Sounds great, but the devil is in the detail and the big problem is that we have very little idea of how to *get* humans to Mars. It's not just that we don't have the technology, which we'll discuss in the following chapter, it's also because there are some huge knowledge gaps in our ability to keep astronauts alive and protect them from the deep space environment. Huge. You see, before you can run you need to walk and right now, when it comes to keeping the crew healthy, we're barely crawling. Deep space is a lethal environment and it's not the sort of place you want to be sending complicated, highly tuned systems such as the human body without lots and *lots* of protection. The Mars One outbound trip will take about seven months and its crew must be kept healthy not only during the transit phase, but also while they're on the surface. We're talking years, but the record for the longest time spent in space is just a little over *one* year. That's the limit of our understanding.

During their mission, the Mars One contestants will suffer radiation exposure, cataracts, vision impairment, bone loss, muscle atrophy, psychological problems and a whole host of nasty medical challenges. If you happen to be on the International Space Station (ISS) and there is a problem, you can simply hitch a ride home on one of the Soyuz lifeboats. But our Mars One group will have no way to abort the mission, except flushing themselves out of the airlock. Mars One says it will train its crew in analogs, but our experience in simulated environments is far from rosy. Ask the Biosphere-2 crew. Or the Mars500 crew. These crews were locked inside simulated environments for more than a year and suffered boredom, lethargy, cognitive deficits and depression. Then there's the problem of Martian dust. For all we know, this stuff is carcinogenic and may produce allergic reactions and pulmonary problems. After all, the Apollo astronauts had similar problems when dealing with lunar dust and developed lunar hay fever as a result. In short, there is a hydra-headed monster of medical problems that need to be resolved before Mars One – or any other manned Mars mission – can journey to the Red Planet. You can gaze longingly at all the CGI visions of sleek spaceships depositing their human payload on the surface of Mars until the cows come home, but the reality is that any such mission must be grounded in facts and analysis. Fantasies do not get you to Mars, no matter how well Mars One spins their case. This chapter explains just how delusional – and I'm not using that term lightly – such a mission is, based on how little we know about the medical mission-killers.

Figure 6.2   Solar flare. Credit: NASA

RADIATION

"Findings by an instrument aboard the Mars transit vehicle that carried the *Curiosity* rover show that radiation exposure for a mission of permanent settlement will be well within space agencies' astronaut career limits. A study published in the journal *Science*, in May 2013, calculates 662 ± 108 millisieverts (mSv) of radiation exposure for a 360-day return trip, as measured by the Radiation Assessment Detector (RAD). The 210-day journey Mars One settlers will take amounts to radiation exposure of 386 ± 63 mSv. This exposure is below the upper limits of accepted standards for an astronaut career. The astronauts should expect one SPE every two months on average and a total of three or four during their entire trip, with each one usually lasting not more than a couple of days. Radiation exposure on the surface is 30 μSv (microsieverts) per hour during solar minimum (Figure 6.2); during solar maximum, dosage equivalent of this exposure is reduced by the factor two. If the settlers spend on average three hours every three days outside the habitat, their individual exposure adds up to 11 mSv per year. The 210-day trip results in radiation exposure of the crew of 386 ± 61 mSv [discrepancy on original website excerpt]. On the surface, they will be exposed to about 11 mSv per year during their excursions on the surface of Mars. This means that the settlers will be able to spend about sixty years on Mars before reaching their career limit, with respect to ESA standards."

Excerpt from the Mars One website.

As with so many mission details, this is one that Mars One seems not only to have overlooked, but dismissed entirely. This is surprising, because if there is one medical mission-killer that stands head and shoulders above all the rest it is radiation. Space is full of the stuff and there is no escape from it. Imagine standing in the desert with a sand storm raging around you, particles of sand stinging your skin. That's ionizing radiation for you, except that radiation won't bounce off your skin like a grain of sand. Oh no: radiation slices right *through* your body, causing catastrophic damage in its wake. And as we shall see, spending a lot of time in an environment like this is as good as a death sentence.

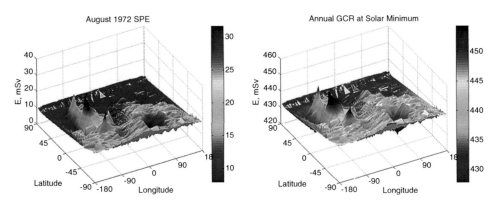

Figure 6.3   Radiation exposure during SPE and GCR. Credit: NASA

The type of radiation that causes most concern for flight surgeons and mission planners is galactic cosmic radiation. Also known as Galactic Cosmic Rays (GCR), this stuff is composed of bits of atoms ripped off following supernovae. Traveling at close to the speed of light, this radiation has the potential to cause terrible damage. Fortunately for those on board the International Space Station (ISS), our planet has a magnetic field that protects astronauts from the worst of this type of radiation, although crewmembers still take a radiation hit during their missions. For the rest of us, we are hardly affected by GCRs here on the ground thanks to our atmosphere, which acts like a shield. That's because as these particles zip through the atmosphere, they collide with atoms and begin to break apart. After this process is repeated many times, the particles gradually lose all their energy and they become weaker. Eventually, the particles are weakened to such a point that they are hardly able to do any damage at all.

But up there in space it's a different story. While astronauts in low Earth orbit (LEO) benefit from the protective cocoon of our magnetic field, they still have to be careful about radiation exposure because radiation damage is cumulative. Arguments are routinely put forward by the Mars fanatics that we can survive the radiation exposure during a six-month trip to the Red Planet because we have had astronauts spend this amount of time in LEO and they have survived. True, but it is impossible to rack up a Mars-equivalent radiation risk during a six-month stint on the ISS, because the ISS is not in deep space and, as

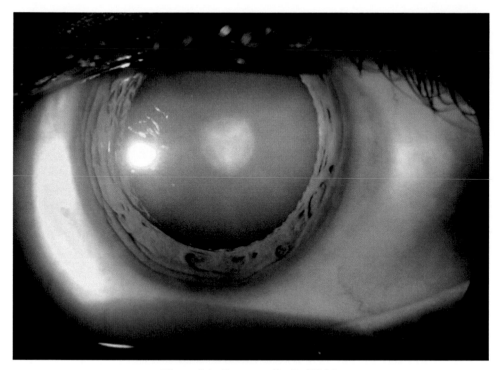

Figure 6.4   Cataracts. Credit: NASA

we shall see, in deep space bad things happen. But before we venture into deep space, what about the damage inflicted on astronauts who spend time on the ISS. Let's begin with cataracts (Figure 6.4).

CATARACTS

Those who have cataracts have degraded vision due to changes in the lens of the eye that make it more opaque, as depicted in Figure 6.4. Over the years, scientists have examined data in NASA's Longitudinal Study of Astronaut Health (LSAH) and found that crew-members who had spent the longest time in space were more likely to develop cataracts than those astronauts who had spent less time in space. Why? Exposure to radiation. The issue of cataracts was first raised during the Apollo era, when astronauts reported seeing bright flashes. The flashes were not some astronomical phenomena, but radiation tearing through the astronaut's eyes like tiny projectiles, leaving a trail of destruction in their wake. As you can imagine, radiation zipping through your eye is not good for long-term ophthalmological health, as evidenced by the fact that several astronauts who flew high radiation missions later developed cataracts just four or five years after their mission.

Cataracts

For you to enjoy clear vision, the lens of your eye must also be clear, because it is this structure that focuses incoming light onto the retina. But if the lens is damaged in any way, clouding will occur resulting in a reduction in clarity, and it is this change that is known as a cataract. Now your body is not only extremely adaptable and highly resilient, it also happens to be very good at repairing and maintaining itself. So, when the lens is damaged, or if new cells need to be manufactured, new fiber cells are created to replace the old or damaged cells. But this process can be upset if the radiation dose is too high. In this case, that metamorphosis is disrupted and the damage becomes permanent. The problem with cataracts is that they take several years to develop and scientists don't know if higher radiation doses accelerate the onset. Either way, cataracts are bad news and a surefire mission-killer for Mars One, or any manned Mars mission for that matter.

CANCER

Now let's move on to the next mission-killer, and it's a big one. The big 'C', no less. Expose humans to radiation and there is an increased risk of cancer. We've known this for years because we have studied the epidemiological evidence from Hiroshima and Nagasaki, although the type of radiation unleashed in those nuclear events is different from the stuff we find in deep space. We're talking about GCR again, and particularly particles known as HZE ions. These HZE ions are deadly because they are very heavy, they move very, *very* fast, and they cause all sorts of damage. How do we know this? Before and after every spaceflight, flight surgeons take blood samples from each crewmember for post-flight analysis. While the astronauts are in space, some of the blood taken in the sample is exposed to gamma rays, which is the sort of radiation we are exposed to on Earth. Then, when the crews return, another blood sample is taken and the scientists compare the gamma ray samples with the blood taken post-flight. Typically, the increase in the amount of radiation exposure is two to three times higher than if the astronaut had stayed on Earth. And don't forget, this increase occurs while the crews are afforded the protection of the Earth's magnetic field. Making matters more difficult for the scientists whose job it is to analyze radiation exposure are all the myriad variables – mission length, crew age, crew gender, shielding, pharmacological countermeasures – which influence the damage that radiation inflicts on astronauts in space. And even if scientists did have a way of assessing the impact of each of those variables, there is still a huge void of knowledge in relation to predicting Solar Particle Events (SPE), cell regulation and tissue responses to deep space radiation, genetic repair in response to radiation, and individual radiation sensitivity factors. Does Mars One have an answer to all these unknowns? Of course not. Just like they have no idea how to characterize the GCR environment en route to the Red Planet and just like they have no idea of how radiation that penetrates shielding affects tissues. And what about the effects that long term exposure to radiation has on immune dysfunction? The list goes on. And on.

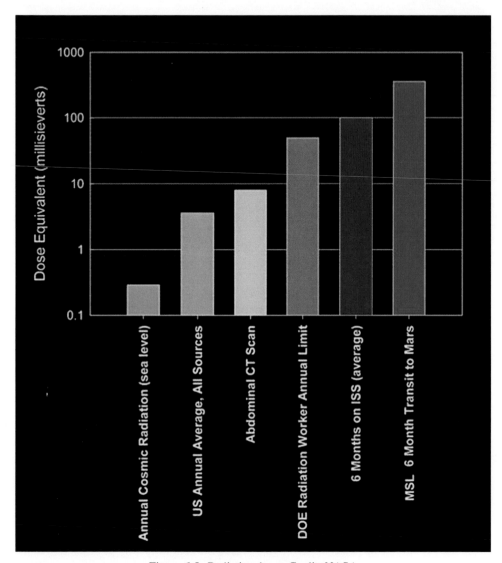

Figure 6.5 Radiation doses. Credit: NASA

But let's get back to cancer. We know a little about how radiation causes cancer by studying populations exposed to radiation following such events as Hiroshima and Chernobyl. These studies show cancer morbidity and cancer risks for 12 tissues and also show evidence for various types of leukemia and tumors of the lungs, bladder and liver. Basically, we know that exposure to a large amount of radiation in a short period of time causes cancer. To give you an idea of comparative radiation doses, Figure 6.5 shows the difference in radiation exposure between terrestrial and the amount accrued during a Mars transit. So how much is too much? After all, NASA sends its astronauts to the ISS, which orbits at an altitude where crews are exposed to a fair amount of radiation. One option is unlimited risk, and it seems this is the approach Mars One has adopted. Of course this policy[1] wouldn't fly with NASA, or any other space agency for that matter, so another approach is to compare ground-based worker limits and use these as a reference point for astronauts working on orbit.

Based on the information gleaned from studying terrestrial radiation workers (such as those who work in nuclear power plants, for example), NASA has decided that an astronaut's radiation exposure limit should not exceed a three percent increase in the risk of radiation-induced death from fatal cancer during the course of their career. This limit has been applied to all LEO missions, as well as any lunar mission that is less than 180 days long. Having said that, NASA also adopts the policy of ensuring that astronauts are exposed to the lowest amount of radiation possible, a legal requirement known as ALARA, or As Low As Reasonably Achievable. Adhering to ALARA is especially important, because cancer risk due to radiation exposure is dependent upon age, gender, sensitivities in tissue types (Table 6.1), and individual resistance.

Table 6.1 Career Radiation Limits

Organ	Tissue Weighting Factor	Organ	Tissue Weighting Factor
Gonads	0.20	Breast	0.05
Red Bone Marrow	0.12	Liver	0.05
Colon	0.12	Esophagus	0.05
Lungs	0.12	Thyroid	0.05
Stomach	0.05	Skin	0.01

To determine the risk of cancer, space agencies use radiation dosimetry and apply physics. First, the body is categorized into tissue types, with each assigned a weighting factor depending on its sensitivity to the risk of cancer. Once the absorbed dose delivered to each tissue type is known (this is derived from dosimetry), the mission duration is inputted and, using some fancy formulae, the risk of radiation-induced death is scaled against Japanese survivor data. If astronauts have been exposed to an SPE event, there are calculations[2] that allow consideration of the orientation of the crewmember relative to the vehicle shielding during the SPE.

So, before Mars One heads off to the Red Planet, they should really pay attention to the risk of cancer from radiation. This will be difficult unless they adopt the unlimited risk policy, because operational parameters for cancer risk have yet to be optimized for Exploration Class missions, and mitigation measures (shielding, pharmacological intervention, which are discussed in the next chapter) are in their infancy. A mission such as Mars One is extraordinarily risky because of the very high probability of vehicle and/or

---

[1] If you are interested in the details of cancer and radiation risk, then you might find NCRP Report No. 153 interesting. This report goes into some detail about how radiation recommendations for beyond LEO are made. I wonder if anyone in the Mars One organization has read it?

[2] Again, if you are interested in the details, the methods used can be found in NCRP Report 126.

life support system failure, and heading off without adequate protection against radiation just increases that risk to levels that are completely unacceptable. Now you may argue that you have to accept these risks, which is fine, but the level of that risk must align with the stakeholders' risk tolerance. Nowadays, a goal of less than one percent mission failure risk is applied when designing the life support and crew systems, and a three percent risk of suffering radiation-induced cancer is applied to the crew for the mission. But we're talking about missions in LEO. What happens when we move out into deep space?

Some of what we know about deep space radiation and the damage it causes is derived from biophysical models, and what these models show isn't encouraging for the Mars One crew. Let's begin with genetic damage. Now your genetic material is very resilient and very good at self-repair, but it can only do so much. And if you put yourself in an environment where lethal radiation sleets through your body continuously, then permanent damage is inevitable. This damage manifests itself as DNA lesions and complex DNA breaks – double and single strand – that occur in clusters. Repair proteins go into action to try to limit the damage, but on a long duration mission these proteins will quickly be overwhelmed. We know this thanks to studies that have used live cell imaging to examine cells hit by heavy ions. The results aren't pretty. Over time, this onslaught of radiation results in mutations because DNA material is mis-repaired. Chromosomal aberrations and chromosomal rearrangements induced by these heavy ions ultimately lead to cell death. Those mutated cells that survive may cause tumors to form, with the lungs and stomach being the most likely sites. Like I said, the outcome ain't pretty. Remember, we know this from modeling, which is a very limited platform of knowledge. We have no idea exactly how malignancies will develop, just like we don't know the mechanisms that cause genomic instability. We have no idea to what degree HZE will cause oxidative damage, or how these nuclei will modify matrix remodeling responses, or to what degree chromosomal rearrangements induced by HZE ions are transmitted to the progeny of surviving cells. We have no idea… well, you get the point: we *really have* no idea! And lately, the prospects for any potential trip to the Red Planet got worse after scientists looked at the data sent back by *Curiosity*, the rover that was sent to Mars in 2011. Strapped to the side of *Curiosity* was a coffee-maker-sized Radiation Assessment Detector (RAD) that measured radiation en route. During the vehicle's 253-day trip, the RAD absorbed about half a sievert.[3] So what happens if one or more of the Mars One contestants are diagnosed with cancer. Rescue? Not likely. It's difficult even to do that in remote locations on Earth.

The case of Jerri Nielsen Fitzgerald

Those of you who follow what happens in the Arctic and Antarctic may argue that humans are capable of pulling off extraordinary rescue missions. One such case that hit the headlines not so long ago was that of a doctor who was diagnosed with cancer while on duty in the Antarctic. Some of you may remember this story. In the late 1990s, Dr. Jerri Fitzgerald

---

[3] Radiation dosage is measured in Sieverts. On Earth, a person absorbs less than one thousandth of a Sievert per year. A cumulative dose of one Sievert might increase the risk of fatal cancer by almost 25 percent. This level of risk would exceed NASA's acceptable risk limits.

was going through a nasty divorce when she noticed a wanted ad in a medical journal for a doctor at the Amundsen-Scott research station at the South Pole. Buoyed by the thought of escaping all the stress of dealing with her husband, Dr. Fitzgerald applied for the position. After a thorough vetting she was offered the job, and arrived at the South Pole in early 1999.

In late May, Dr. Fitzgerald noticed a lump in her right breast, which prompted her to contact Dr. Kathy Miller, an oncologist who prescribed treatment via email. Since Dr. Fitzgerald was the only crewmember with medical training, she had to be helped by her untrained co-workers. One such co-worker was a welder, who performed a biopsy on Dr. Fitzgerald, while a maintenance worker prepped slides for transmitting the results. Later in the season, chemotherapy and other medical equipment was airdropped to the base in an effort to help Dr. Fitzgerald as much as possible. But the chemotherapy weakened the doctor and her condition remained life threatening. Finally, in October, a Hercules aircraft landed at the pole and evacuated the stricken doctor. Two years later, Dr. Fitzgerald wrote an account of her ordeal in '*Ice Bound: A Doctor's Incredible Battle for Survival at the South Pole*'. Required reading for anyone contemplating a one-way mission to Mars. Dr. Fitzgerald's cancer went into remission but recurred several years later, causing her death in 2009.

Why mention the case of Dr. Fitzgerald? Because even in a well-equipped base such as the Amundsen-Scott base, Dr. Fitzgerald's condition quickly exceeded the on-site medical capabilities. So what hope does a crewmember who contracts cancer have on the surface of Mars? No chance. Because no intervention will be possible. No Hercules aircraft can come to the rescue and no amount of computer-driven diagnostics, robot surgeons or telerobotic surgery will make a damn bit of difference. The stark reality of sending Kendra, Simon, Josh and Selena to the surface of Mars is that these crewmembers will be doomed, for the simple reason that it will be impossible to carry sufficient medical equipment and supplies to allow the crew to deal with anything beyond the most benign contingency.

Another spanner in the works for the Mars One organization is the composition of its crew, half of which will be female. This won't work. At least, not if NASA standards are applied. That's because women have a lower threshold for radiation exposure than men. Just check out the mission durations of female astronauts if you don't believe me. The fact of the matter is that women fly only about fifty percent of the missions their male counterparts fly. So, if you happen to have two X chromosomes, you're less likely to fly long duration missions. And if you happen to be a female Mars One contestant? Well, I guess you'll be the first to go into the body-bag.

In short, every cell nucleus in the body of each Mars One contestant traveling to the Red Planet will be traversed by a proton or secondary electron every few days, and by a HZE ion every three or four weeks. This will result in complex genetic damage, the mis-joining of DNA strands, DNA deletions, chromosomal aberrations and chromosomal rearrangements that will eventually lead to cell death. Chromosomal instability will occur thanks to telomere deletions, micro-lesion formation will occur, as will accelerated striated aging, altered dopamine function and neurodegeneration. Ultimately, all this radiation-induced disruption will induce cancers. Who will look after the Mars One contestants then? For those with a morbid fascination for watching people die, this will be reality television at its very best. Millions will tune in on their high definition televisions to watch Mars One contestants vomit their intestinal linings and die the agonizing deaths that only cancer can deliver. Ratings will spike, but once the vehicle is full of radiation-soaked corpses, what then? Ask Bas.

BONE DEMINERALIZATION

"When Mars One astronauts arrive on Mars (62% less gravity than Earth), they would theoretically be stronger compared to an astronaut returning to Earth's gravity after a mission of similar duration. A recent study of ISS astronauts, with mission durations ranging from 4-6 months, showed a maximum loss of 30% muscle performance (and maximum loss of 15% muscle mass). However, we intend even to lower these numbers. With recent and emerging scientific research of effective long-duration countermeasures, Mars One will take advantage of the ~10 years prior to the launch of the first colonization mission to observe and select the most suitable astronauts and countermeasures to ensure a safe and successful mission. Astronauts will suffer a loss of bone density; however, the problem can be mitigated with appropriate and well-designed countermeasures, including but not limited to exercise and pharmaceuticals. Continued research and advancements in this area will surely produce even more effective countermeasures within the ~10 year period of preparation prior to the first Mars One launch. Once on the Mars surface, astronauts will be able to take advantage of the force of gravity (albeit, less than that of Earth) to assist them in the reconditioning and adaptation process, which will result in bone remodeling that will help to strengthen the astronauts' bones."

Excerpt from the Mars One website.

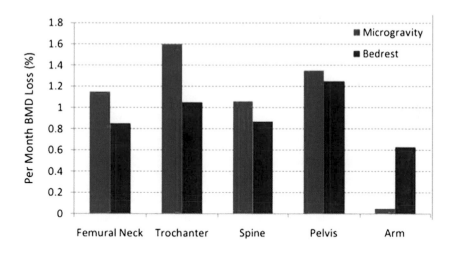

Figure 6.6   Short term (< 6 months) space mission and bed-rest induced loss of bone mineral density, averaged per month loss. Data adapted from reference LeBlanc et al. (LeBlanc et al. 2000b; LeBlanc et al. 2007)

For decades, we have known about the physiological changes that occur when astronauts spend extended time in space. For decades, scientists have observed changes in bone mass (Figure 6.6), measured decrements in aerobic capacity and quantified losses in bone density. You would think that after all this time we would have a good idea of how to mitigate these losses, but we don't. Sure, scientists have developed countermeasures, devised fancy exercise systems and concocted highly specific nutritional and pharmacological intervention strategies in an effort to keep crews as healthy as possible. But the absence of gravity is a difficult nut to crack when it comes to keeping astronauts healthy. Without any gravity acting on the bones, astronauts lose bone density. That's because without the static loading we enjoy on Earth, bones weaken as a result of bone demineralization, with the greatest bone loss in the weight-bearing bones such as the lower extremities (the lumbar vertebrae, the femur, and the tibia for example). On average, about one to two percent of bone is lost per month in these bones. Per month! As you can imagine, losing so much bone puts astronauts at a very high risk of fracture post-mission. But for our Mars One contestants, there will *be* no post-mission. Take a look at Figure 6.7. That's a photo of Chris Hadfield after his mission to the ISS. That mission lasted for 120 days and when he returned, he was carried away on a couch. Who will be waiting for the Mars One crew? Oh, but I forgot, the Mars One organization will have solved the bone loss problem in the next few years! What a relief.

Figure 6.7 Chris Hadfield following his tour on board the International Space Station. This recovery option will not be available on Mars. Credit: Canadian Space Agency

But let's suppose Mars One isn't able to solve a problem that has remained unsolvable for decades, despite the efforts of the very best scientists on the planet. What then? Well, then our Mars One contestants are in for a world of hurt. Let me explain. You see, it isn't just bone loss that long duration astronauts have to worry about. As soon as astronauts enter space, their bodies start excreting calcium. Lots of it. In fact, a 60 to 70 percent increase in urinary calcium is detected in the first couple of days on orbit, and this continues all the way through the mission. This is a big problem, because it increases the risk of astronauts developing kidney stones. Now some of you may have been unfortunate enough to experience the excruciating pain that accompanies a kidney stone, or perhaps you know someone who has suffered kidney stones, but in case you haven't, here's a synopsis.

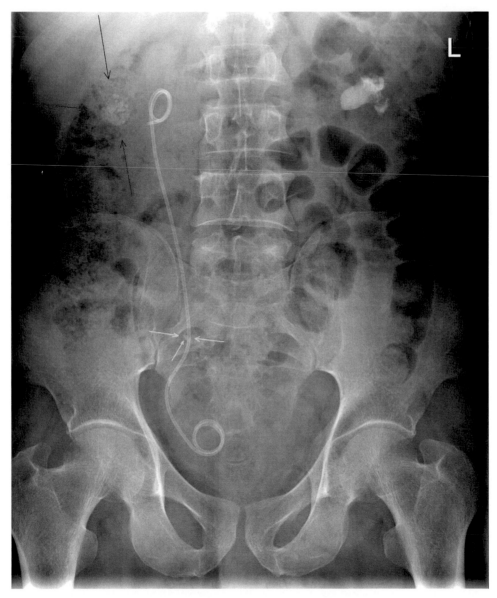

Figure 6.8 Kidney stone. Credit: NASA

Let's begin with the kidney stone. This is a hard mineral material that forms when there is an excess of stone-forming material, such as calcium, in the urine. Symptoms include blood in the urine and pain. Lots and *lots* of pain. This pain may be accompanied by nausea and/or vomiting, fever and chills, and frequent urination. What would happen to a Mars One contestant who suffered a kidney stone? Well, the medical officer (which could be the one suffering the kidney stone) would break out the pain-control medication, prescribe lots and lots of fluid and probably inject Ketorolac, which is an anti-inflammatory. If the crewmember is suffering nausea and vomiting, intravenous pain medications will also be prescribed. If the crewmember is lucky, the stone will pass within 48 hours after treatment, but it depends on the size of the stone. Stones less than 4 mm in diameter will pass 80 percent of the time, while stones that are 5 mm in diameter will pass 20 percent of the time. But if the stone is 9 mm in diameter or greater, then more drugs (Adalat, Procardia, Afeditab) may be needed. And if that doesn't work then surgery – Extracorporeal Shock Wave Lithotripsy (ESWL) – is one of the final options (see sidebar). Lithotripsy is a procedure that uses shock waves to break the kidney stone into small pieces. On Earth, the procedure is fairly standard, but the chances of the Mars One vehicle having the equipment to perform this procedure are slim to none. But let's pretend the medical kit has the tools to perform the lithotripsy, but the procedure fails. What then? Well, now the crewmember is facing surgery – percutaneous nephrolithotomy – a procedure which involves slicing through the skin using specialized instruments inserted through a small incision in the patient's back. The chances of any crewmember having been trained to do this, or being capable of being trained to do this, are non-existent.

Dealing with kidney stones

A lot can happen in ten years in the medical technology arena. One of the pieces of kit that might have been developed to a size that fits into the Mars One medical kit is an ultrasound technology that can take images of the kidney and detect kidney stones. The company working on this device is Sonomotion, and its non-invasive solution for dealing with kidney stones uses ultrasound waves to reposition kidney stones before breaking up the offending item using a technology called Break Wave. It could be a Godsend for a crewmember stricken with kidney stones. Now all that has to be done is devise a way to protect our Mars One crew from radiation, blindness, bone loss, cataracts, etc., etc.

So, kidney stones will be a risk, but let's go back to the problem of bone loss. How do we deal with this problem today? Well, flight surgeons use Dual-energy X-ray Absorptiometry (DXA) to measure bone density pre- and post-mission. Obviously this won't be available during a Mars One mission because of the bulk of the equipment, so none of the contestants will know how much bone density they will have lost. And what about countermeasures? Most countermeasures are exercise-oriented and involve applying resistance. This, combined with vitamin (D and K) and calcium supplements help astronauts protect their skeletal system, but it isn't enough. Not by a long shot. Despite the assurances advertised on the Mars One website, today's astronauts don't expect to recover their bone density until at least three years post-mission and there are some astronauts who never recover their pre-mission bone density. Ever! That doesn't bode well for Mars One does it?

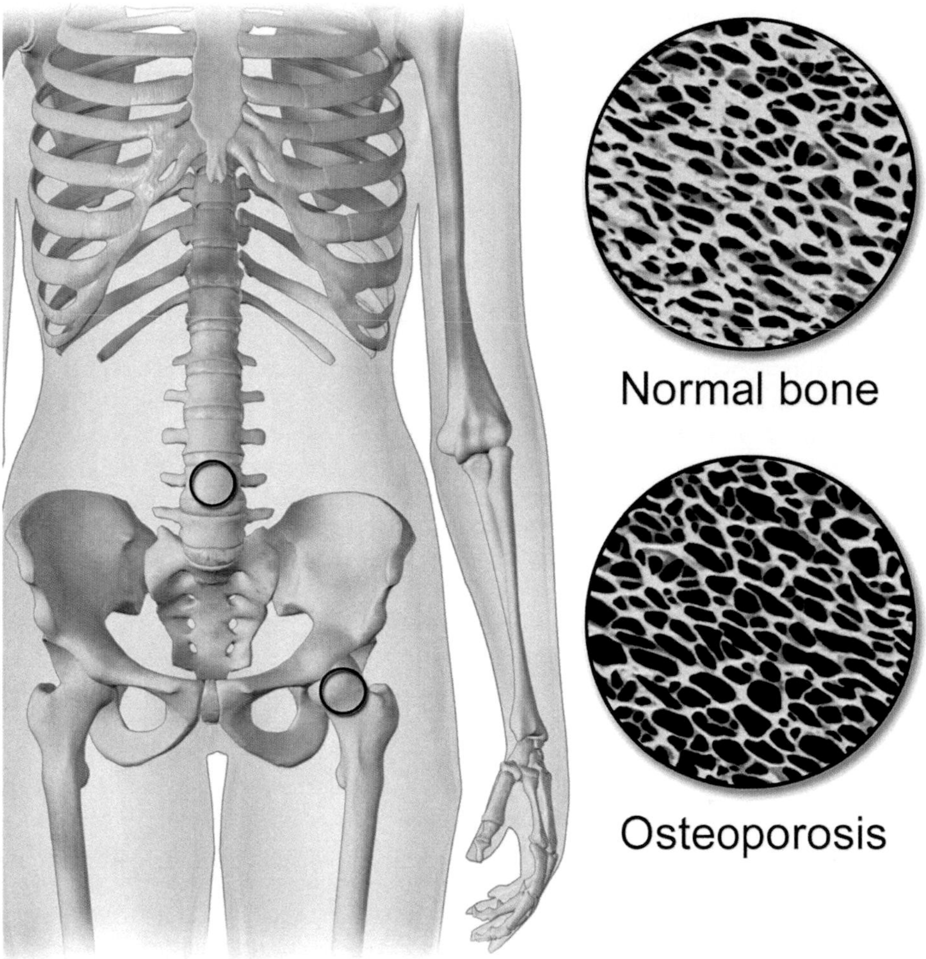

Figure 6.9 Osteoporosis: The upper image shows strong bone and the lower image shows what bone might look like after a few months en route to Mars. Mars One take note. Credit: Bruce Blaus

Take a look at Figure 6.9. At the top is normal bone and at the bottom is osteoporotic bone. It looks bad doesn't it? Well, for a Mars One astronaut this could be a death sentence, and here's why. As bone loses density, it undergoes significant pathological changes, some of which are changes to the bone architecture. And as a result of these pathological changes, bone that has undergone microgravity-induced deterioration has an increased risk of fracture that is five times that of normal bone. Now let's get back to our Mars One crew. It's the first day on the surface of Mars (we'll make the highly unlikely assumption that at least some of the crew have survived the trip) and millions of reality television viewers are tuned in to witness this momentous event. Selena takes her first step on the Red Planet and her leg snaps! Now the crew is in a bit of a pickle, because the Mars One organization was so confident it had solved the bone density issue that it hadn't planned for such a contingency. After all, the Mars One mantra on the subject of bone loss is: "Mars One astronauts arriving on Mars will theoretically be stronger compared to an astronaut returning to Earth's gravity after a mission of similar duration." For those of you who have been unfortunate enough to break a bone, you will know about all the many potential problems and complications associated with a fracture. But for those of you who have been lucky enough to avoid breaking bones, here is a synopsis of what happens here on Earth.

First, the type and location of the fracture will determine the treatment. Simple fractures such as transverse fractures, in which the break is straight across the shaft, are relatively easy to treat, whereas complex fractures such as a comminuted fracture, in which the bone is broken in three or more pieces, require more complex treatment. But Selena has an *open fracture*, which is characterized by bone fragments sticking out through the skin. This is about as bad as fractures get. With no emergency doctor on hand, the designated medical officer will have to put the bones back into alignment, a procedure called *reduction*. On Earth, such a procedure would most likely be performed by an orthopedic doctor, but there won't be many of these running around on Mars. Too bad. If the medical officer somehow manages to align the bones, after somehow managing to manipulate the bones inside the skin, he or she will then have to figure out how to insert all the pins, screws and/or metal plates needed to hold the bones together. On Earth, such a procedure would be performed with the aid of X-rays and Computed Tomography (CT), but even with all these diagnostic tests, the outcome of surgery for an open fracture is far from certain. That's because an open fracture needs a lot of treatment. You see, the chances are that the soft tissues around Selena's fracture will be damaged, which will mean it will be a while before she can have surgery. In this case, the best option will be to apply a temporary external fixator, comprising pins and screws that are inserted into the middle of the bone, with the pins and screws attached externally (see sidebar). This stabilizes the bones until Selena is ready for surgery. But who are we kidding? In reality, if this scenario was to play out, Selena would be as good as dead. And even if, miraculously, the crew managed to pull through and stabilize Selena, they would still have to operate at some point. How would they perform intramedullary nailing and how would they reposition the plates and screws? Remember, this all requires surgery, and who would be qualified to do this? And in a reduced gravity environment, to boot!

External Fixation

As you can imagine, this procedure may result in all sorts of problems. The site around the pins could become infected, or the pins could loosen and/or break. And there's a good chance neurovascular damage could occur at the site of pin placement, or the bones could be misaligned due to poor pin placement.

Compounding the problem would be Selena's bone quality. Remember, she has just spent several months in zero G, so her bone quality won't be tip top. This will mean her fracture will be slow to heal and may require a bone graft or the use of artificial bone fillers. What are the chances of these options being available on the Red Planet? None. Exactly. And how would the crew prevent infection and deal with the blood loss during surgery? And even if, by some series of divine interventions, Selena is stabilized, her recovery will take the best part of a year and will require the attention of at least one other crewmember to assist in the rehabilitation. Which crewmember will be nominated as the physical therapist I wonder? And during this year of rehabilitation, how will the crew respond to the infections? Remember, this is an open fracture, which is particularly prone to infection, which means Selena will require long-term intravenous antibiotic treatment as well as numerous surgeries to clear out the infection sites. Not gonna happen! In reality, the crew and the vehicle's medical resources would be rapidly overwhelmed. There is no way the crew would be able to deal with the myriad immediate complications – excessive bleeding, soft tissue compromise, neurovascular injury, take your pick; or the longer term issues – delayed union, non-union, avascular necrosis, osteomyelitis, take your pick again; or the possible systemic issues – gangrene, septicemia. Even if Selena was lucky enough to suffer a simple fracture, the chances of her recovering fully are slim, because in a reduced gravity environment the process of bone remodeling would be severely compromised. For the rest of the mission, Selena would be no better than an invalid. Perhaps they should take a wheelchair along with them. Just in case. Your call Bas.

I know a lot of flight surgeons and a number of doctors and I've asked them for their perspective on this scenario. After a roll of their eyes, their answer is always the same: Selena wouldn't stand a chance. No way.

Scientists have known about the bone density problem for some time (see sidebar), which is why there has been a concerted effort over the years to try to develop countermeasures. Exercise has been tried with varying degrees of success, as have biphosphonates, which have been tested on board the ISS since 2008. Biphosphonates help block the breakdown of bone, but to date this countermeasure has been insufficient to stem the tide of bone loss. So where does that leave Mars One? Well, there is always a chance that a magic bullet will be found that will stop bone loss, but that is a long shot. Alternatively, Mars One could screen for those candidates who have high bone density, but remember bone density isn't the whole story, because there is still that matter of bone architecture to deal with and we still are very much in the dark on that subject. The bottom line is that a six-month mission requires years of rehabilitation and there are no rehabilitation specialists living on Mars. There are also huge knowledge gaps regarding how already weakened bones deal with a reduced gravity environment, how fractures heal in reduced gravity, and the risk to astronauts' bone health following years living in reduced gravity.

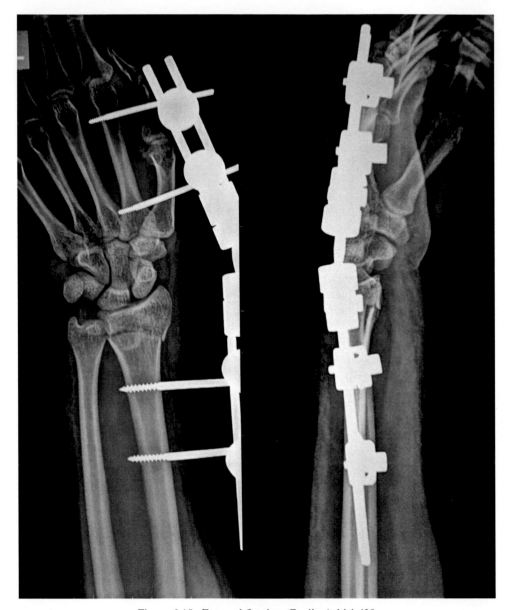

Figure 6.10   External fixation. Credit: Ashish j29

Bone loss

Scientists reckon astronauts lose bone density because the *osteoblasts* (bone form-
ing cells) don't form as fast in an environment – space – where there is no gravity.
Making matters worse is the way lack of gravity affects the action of *osteoclasts*,
whose job it is to break down bone minerals: in space, these bone resorption cells
work at a much faster rate, which means bone just can't repair itself.

But let's really, *really* stretch credulity and imagine a Mars One mission is given the go ahead. Based on what we know from decades of research, what might happen to the bones of the crew? We'll assume a six-month flight to Mars and that the rates of bone loss are the same as the average rates of loss observed in long duration missions on board the ISS. What happens once the crew is on the surface is speculative (we'll further assume that radiation doesn't kill them), but we have a reasonable idea thanks to computer modeling and scaling the loss that occurs on orbit. We will also assume that the crew exercises regularly and takes the pharmacologic supplements that are used currently. On the outbound trip, the crew will lose bone density, bone strength, muscle mass and muscle strength. The strength of their femurs will be reduced by about 12 percent and their muscle strength will be decreased by about ten to 15 percent. Assuming they land safely, the crew will be living in an environment where gravity is 38 percent that of Earth gravity. It might just be possible that the crew will be able to function in this environment, despite the loss of bone density suffered in the first six months of the mission. Might. What is certain is that no matter how much they exercise, the crew will continue to suffer losses in bone and muscle strength as a result of the lack of gravity. In fact, based on computer modeling, the Mars One crew can expect to lose an additional nine percent of their bone strength in the first year alone. That's bad enough, but the news gets worse because the crew will also continue to lose muscle strength – an additional 18 percent. Admittedly, these numbers do not consider the effect of wearing spacesuits which may attenuate some losses, but this effect is immaterial because the crew won't be spending any time on the surface on account of the radiation. Instead, they will be holed up living in bunkers deep underground. As the mission continues, the crew will continue to lose bone and muscle strength and their fracture risk will steadily increase. If the crew decides to spend time on the surface – ill-advised on account of the danger from radiation, but this is a reality television show, so the producers will probably want footage of the crew on the surface – then their bone loss will be compounded by the radiation exposure.

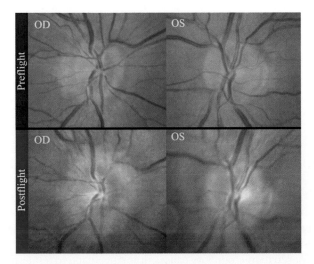

Figure 6.11   Visual Impairment Intracranial Pressure. Credit: NASA

GOING BLIND

Cataracts. Cancer. Bone loss. That's three surefire mission-killers right there, but I haven't finished. The next major problem that prevents any serious attempt at a manned Mars mission is an eye condition that blurs eyesight. I'm talking about Visual Impairment Intracranial Pressure (VIIP - Figure 6.11). Now, I happen to know a little about this eye condition because I know the NASA lead personally and I also wrote the book on the subject.

> "We are certainly treating this with a great deal of respect. This [eye condition] is comparable to the other risks like bone demineralization [loss] and radiation that we have to consider... It does have the potential for causing mission impact."
>
> *Dr. Rich Williams, NASA's Chief Health and Medical Officer*

The VIIP headache is a recent Mars mission-killer, but potentially just as deadly as cancer or fractures. The condition mainly afflicts astronauts who spend long periods of time in space. When the issue was raised by an astronaut in 2007, NASA decided to survey the health records of 300 astronauts, and found that about 30 percent of those who had flown Shuttle missions and about 60 percent of those who had flown longer ISS missions had reported blurring of their eyesight. The revelation spurred the media to write all sorts of sensationalistic articles, hinting that astronauts who spend too long in space could go blind. While going blind is an extreme possibility, just having astronauts suffering from blurred vision is unlikely to result in a positive mission outcome. After reviewing the medical records of the astronauts who had suffered VIIP, scientists characterized the problem as being similar to an Earth-bound condition termed *papilledema* which is caused by increased intracranial pressure (ICP). In space, this is caused by an increase in cerebrospinal fluid (CSF). The problem with VIIP is that there is no rhyme or reason to the symptoms. Scientists just can't find a smoking gun. Some astronauts experience visual deficits post-mission while others don't. Some crewmembers suffer the structural changes that accompany VIIP but they suffer no symptoms, and yet other astronauts suffer permanent damage. That's not good news for those contemplating a multi-year mission to distant planets. After all, what's the use of sending a group of astronauts to the Red Planet if half of them are myopic when they arrive? Befuddled contestants bumping into each other on the surface hardly makes for compelling reality television, does it? Then again, perhaps it does. And what if the entire crew goes blind? After all, here on Earth, papilledema can lead to blindness if not treated. The problem with VIIP is that no-one has been up there long enough to determine just how bad it can get. Maybe the Mars One guinea-pigs will be able to answer the question?

DISCOMBOBULATION

Now let's move on to mission-killer number five. Discombobulation. Yes, that is a word: check your dictionary if you don't believe me. But NASA prefers to use the term *orthostatic intolerance*, among others, to describe the re-adaptation that astronauts undergo when returning to Earth after a lengthy stay on orbit. Astronauts become discombobulated because being exposed to microgravity causes dramatic adaptive reinterpretation of visual,

Figure 6.12 Scott Kelly on his return from nearly 12 months in space. Credit: NASA

vestibular and proprioceptive information. And the longer astronauts spend in space, the longer they must spend re-adapting to Earth's gravity. During the first days and weeks back on Earth following a six-month stint on the ISS, astronauts experience problems in their posture, gait and spatial disorientation. Even getting up and going for a walk is a challenge. There have been countless astronauts who have worn their NASA issue diapers (MAGs, for Maximum Absorption Garment) to bed for the first days back on Earth because getting to the washroom from the bed proves too challenging. After many, many years of ISS operations, scientists have determined that astronauts need about two weeks to recover their postural equilibrium control. And that's not a problem, because here on Earth the astronauts are taken care of by rehabilitation specialists. Who will look after the Mars One crew on the Red Planet, I wonder? And what happens if the crew has to respond to a contingency event? You see, postural stability and equilibrium aren't the only sensorimotor changes observed in long duration astronauts. Other skills that take a hit are manual dexterity, speed and accuracy of movement, eye-head coordination, and visual target acquisition. These are all pretty crucial skills for when your spacecraft is heading towards the surface of a planet at a rapid rate of knots and you need to react to an emergency event. Scientists have been researching how microgravity affects sensorimotor skills for decades and the news isn't good for long duration astronauts. In addition to motor skill deficits, crewmembers who spend several months on orbit also experience difficulty in stabilizing their gaze, have trouble steadying their eyes on targets and find that their eye-head coordination is diminished. For an astronaut who is tasked with landing a spacecraft on Mars, such sensorimotor decrements will almost certainly result in a negative mission outcome.

"Anything that can happen to you and me on Earth can happen in space. You can have a kidney stone, a headache that doesn't resolve, or elevated pressure on the brain. You can even have a heart attack. NASA needs to be worried about all the medical repercussions of an unresolved medical problem."

Dorit Donoviel, Deputy Chief Scientist,
National Space Biomedical Research Institute

NSBRI's Deputy Chief Scientist is right. In the big wide vacuum that is space, there is plenty that can go wrong, which is why space agencies have such extraordinarily demanding medical standards. To have any chance of being selected as an astronaut, you need to be in perfect health. Depending on whether you plan on being a pilot or mission specialist, these medical standards vary, but at the very minimum you need to pass an FAA Class III medical or equivalent. Here's an excerpt from §67.303, which describe the eye standards:

(a) Distant visual acuity of 20/40 or better in each eye separately, with or without corrective lenses. If corrective lenses (spectacles or contact lenses) are necessary for 20/40 vision, the person may be eligible only on the condition that corrective lenses are worn while exercising the privileges of an airman certificate.
(b) Near vision of 20/40 or better, Snellen equivalent, at 16 inches (40 cm) in each eye separately, with or without corrective lenses.
(c) Ability to perceive those colors necessary for the safe performance of airman duties.
(d) No acute or chronic pathological condition of either eye or adnexa that interferes with the proper function of an eye, that may reasonably be expected to progress to that degree, or that may reasonably be expected to be aggravated by flying.

But it doesn't matter how thorough the medical screening practices are – those demanding medical standards that space agencies employ are the most rigorous of any organization on the planet – no astronaut is immune to these medical issues. And all the training and all the best flight surgeons in the world can't prevent every type of medical contingency. Those lucky enough to work on board the ISS can take comfort knowing that evacuation is always an option if a medical condition should strike. It would cost millions of dollars, but if an astronaut needed to be evacuated it could be done. En route to Mars is another kettle of fish. Any medical emergency may be a death sentence. And the chances of that happening are high simply because the longer astronauts spend in space, the greater the possibility of a medical emergency arising. Those Mars One contestants will be up there for an awfully long time.

7

Technological feasibility

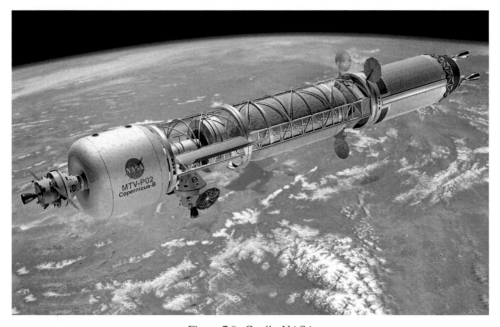

Figure 7.0 Credit: NASA

"I really counsel every single one of the people who is interested in Mars One, whenever they ask me about it, to start asking the hard questions now. I want to see the technical specifications of the vehicle that is orbiting Earth. I want to know: How does a space suit on Mars work? Show me how it is pressurized, and how it is cooled. What's the glove design? None of that stuff can be bought off the rack. It does not exist. You can't just go to SpaceMart and buy those things."

Chris Hadfield, who knows a thing or two about the
manned spaceflight business.

© Springer International Publishing Switzerland 2017
E. Seedhouse, *Mars One*, Springer Praxis Books, DOI 10.1007/978-3-319-44497-0_7

Note: what follows is a dramatized account of an Entry, Descent and Landing (EDL) experienced by our intrepid Mars One contestants (Kendra, Selena, Josh and Simon), who were introduced in Chapter 5.

The spacecraft was dying around him. And it was all because of a mistake he'd made. A mistake he certainly *wouldn't* have made if it hadn't been for the radiation hit three days earlier. In a moment of lucidity, he remembered the brain scan showing the path of the heavy ion that had sleeted its way through his brain, before Selena jolted him back to their reality.

"Deploying supersonic parachutes," Selena announced, punching a button on the flight console.
A muffled bang echoed around the flight deck as the parachutes deployed.
"Mach three point two."
Their speed was too high. Simon tried the reaction control thrusters, to no effect. He tried the flight control inputs. Nothing. They were committed to a ballistic entry. The supersonic chutes would bleed off some of the speed, but the surface was coming up awfully fast.
"Mach two point three and falling. Altitude forty-one thousand. Forty," Selena called out.
"Where are the landing coordinates?" Josh demanded.
"We're still good," Kendra replied. "Bringing up the heading alignment corridor now."
"Altitude thirty-two thousand, four hundred. Mach one point three," Selena said.
The heading alignment corridor hologram snapped into view.
"Calculate trajectory and configure landing mode," Josh said. "Let me know when we've reached subsonic."
Selena nodded. She didn't think they had a hope in hell of pulling this off.

They'd been trained for contingency events in the sims, but nothing in the information vacuum that was The Mars One Handbook had mentioned anything about this.
And they hadn't suffered radiation-induced brain damage in the sims.
She peered through the hole in the cabin at the red surface rushing up at them.
"Deploy the crash shutters. This is gonna get ugly," Josh ordered, as the capsule careened through the tenuous atmosphere.
Simon retched another mouthful of bile and flicked a switch. He'd been taking the radioprotectants, but his radiation hit had been a severe one. The crash shutters slid into place.
"We're shedding big altitude," Selena said.
She vomited another mouthful onto the deck and cursed the side-effects of the amifostine.
"I know. I know," Josh replied.
He knew he hadn't bled off enough speed. They had already dropped below ten thousand feet and they still hadn't reached subsonic speed. There was only one option left.
"Select emergency program number seven," Josh said.
Selena ran through a sequence of switches. On the flight panel, they watched as explosive bolts rapid-fired around the ship, blowing away non-essentials that might compromise what was sure to be a painful crash landing.

"Three thousand feet. Subsonic. Mach point nine," Selena said, incredulous they were still alive.

The subsonic parachutes deployed automatically. She continued to give Josh airspeed, altitude and rate of descent, all the while hoping that all the stories she'd heard about his skills as a pilot were true.

The capsule rolled sickeningly.

Josh nodded. They were coming in nose-high and the ship was rolled to the port side due to the air pressure on the gaping hole in the hull. He exchanged a desperate look with Selena.

"Where can we land this damn thing?" Josh asked.

"Altimeter indicates uneven terrain," Kendra said. "Maximum altitude three hundred feet. The capsule screamed towards the surface, bucking and rolling.

"Thirty seconds to landing," Josh said.

He had half a minute to get the capsule into some sort of landing configuration. At this stage, Josh was no longer maneuvering. He was just searching for a place to crash-land. Behind him, he heard Kendra say a prayer, asking Jesus to forgive her sins. Simon, still moaning, made a quick promise to reform if he walked away.

Josh searched for a flat area of terrain that might serve as a landing area, but every-where was too rugged. He signaled to Simon to initiate retro-propulsion, which shed more speed, but they were still careening along at more than three hundred knots as he lined the capsule up with the surface. Blinding sunlight flared in from the sun as Josh desperately fought to balance the ship on an even keel.

"Five hundred fifty feet and falling fast. Real fast," Selena screamed above the roar of air rushing through the hull breach.

Collision alarms shrilled into action as the huge red mass rose into view.

It was going to be a bumpy landing.

Josh instructed the crew to secure as many loose items as possible. Any small object could become a deadly projectile in a crash landing.

"One hundred. Ninety. Eighty…" Selena's voice was lost in the cacophony of noise generated by alarms and rushing air.

She tightened her straps, said a silent prayer and watched as the rust-colored surface exploded into unwelcome detail.

"Brace yourselves. We're going in," Josh shouted.

The impact was titanic. In an instant, the windshield imploded. Chairs ripped from their moorings. Simon's body slammed into the ceiling. The Mars One capsule was reduced to a vortex of chaos and blurring debris.

The hull began to crack open, splintering like a shell. It was the last thing Selena remembered before blacking out.

Dramatic? Yes. Likely? Most probably. Or worse. Much worse. That's because the technology to land humans on Mars does not exist. You can take as many risks as you like and you can philosophize all sorts of nonsense on a website, but it doesn't change the immutable fact that a manned mission to the Red Planet lies beyond the ragged edge of technological possibility. But of course it depends on your level of risk. Listen, the last time someone was as off the mark as Mars One is, was when British Prime Minister

Neville Chamberlain stepped off an airplane in 1939, waving a piece of paper in the air and reassuring the country that there would be no war with Germany. Not convinced? Here's another nugget from the Mars One website:

> "No new major developments or inventions are needed to make the mission plan a reality. Established suppliers can build each stage of the Mars One mission plan. While most of the components required are not immediately available with the exact specifications, there is no need for radical modifications to the current component designs."

> *Excerpt from the Mars One website, addressing the*
> *technological feasibility of landing humans on Mars.*

LAUNCHERS

See what I mean? Where are the rockets or the capsules? What is the EDL architecture, such as the one dramatized in the excerpt above? Which closed life support system will they be using? Actually, let's leave that last question for the next chapter because that topic deserves extra special attention. For this chapter, let's focus on all the other bits and pieces that are needed to get to Mars but which don't exist today, and likely won't exist for quite a while. We'll begin with the rocket and we'll say this:

There is no rocket in existence today that can send a manned mission to Mars. In fact, there is no rocket that can get humans beyond Earth orbit. None.

> "SLS will be the most powerful rocket in history for deep-space missions, including to an asteroid and ultimately to Mars."

> *NASA, describing its next big rocket*

The Space Launch System, or SLS (Figure 7.1 and 7.2), will be a really big rocket. It may not have a flight manifest, it may have been subject to delays, and the test schedule may be shaky at best, but the SLS is one of only two vehicles being developed (SpaceX's Falcon Heavy being the other candidate) that have any chance of getting humans to Mars. Born out of the long defunct Constellation program, the SLS is something of a salvage operation, with the purpose of finally getting humans somewhere beyond Earth orbit. The vehicle's development costs have been estimated at around $18 billion and when it is finally built, each launch will cost close to one billion dollars. One BILLION!

That figure is more than the $500 to $700 million per launch quoted by NASA, but the devil is in the detail. You see, there are no reliable cost estimates for the SLS, but that is not to say a cost per launch calculation can't be made. Here goes. First of all, it is inconceivable that a monster rocket like the SLS could possibly compete for launch costs with the Shuttle. Part of the reason is that all those very expensive rocket engines and other equipment will either disintegrate in the ocean or burn up in the atmosphere because, unlike the Shuttle, which could be partly refurbished, the SLS is all expendable. So much for progress! The Shuttle, based on average annual program and development figures, cost about one billion dollars per mission. The SLS and Orion combination? Well, let's see shall we? According to NASA, the cost per launch will be around $500 million, but this

Figure 7.1   Space Launch System. A very expensive ride to space. Credit: NASA

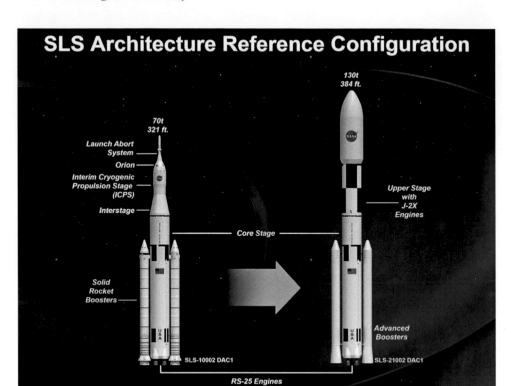

Figure 7.2   Space Launch System showing the options. Credit: NASA

does not factor in development costs, which are estimated to be $30 billion. That $500 million per launch figure also doesn't factor in the annual operating budget, or the fact that none of these solid rocket booster casings can be reused. So let's start crunching the numbers. The development cost of the SLS and the Orion will be around $3 billion per year for at least ten years. At least. If the program runs for 30 years, then the average launch cost will be $1 billion per year. But that number doesn't consider annual operating costs, such as maintaining the small army of workers needed for launch operations. Way back in the days of the Shuttle, the annual operating costs were about three to five billion dollars, but the SLS isn't an orbiter, so that reduces the maintenance costs to around $2 billion a year. Now let's move on to the Orion. This is reusable, but it remains to be seen how many missions it will survive, since landing in saltwater will not be kind to the unpressurized sections. In reality, the capsule will probably have to be disassembled between flights.

Based on external estimates, the SLS-Orion system could be launched once a year. So, with an annual operating cost of $2 billion per year, plus $1 billion for the SLS and another $1 billion for each Orion, we can calculate an operating budget of $4 billion per year. But wait a minute: we haven't factored in the development costs, which are $30 billion over 30 years. Add another billion and we have a number of $5 billion per launch. But that's assuming there are missions every year and we know there aren't. That's because the cost of the SLS and the Orion is so high that NASA cannot develop payloads to fly on the system! The agency can either develop the SLS-Orion, or payloads for it, but not both at the same time. Oh dear.

"We have no experience with a human-rate flight system that only flies every two or three or four years. And I believe that's cause for serious concern. It's not just simply a matter of maintaining program momentum. It's not even purely a matter of efficiency. It's also a matter of keeping the flight team sharp and safe."

NASA Advisory Council Chairman Steven Squyres, in a briefing to members of the House Space, Science and Technology Committee

That little snippet of information, which was revealed in 2013, indicates NASA is planning to fly the SLS only once every four years. This is terrible news when you consider the launch readiness challenges faced by workers tasked with starting up Shuttle operations for return to flight programs after the two accidents. It also dramatically alters our launch calculations, because if the system begins operations in 2023 and goes on for its 30-year lifespan then the SLS could only be launched seven times! This means the cost per launch increases from $5 billion to around $9 billion! But wait, we haven't factored in the development costs: $1 billion for the Orion, plus $1 billion for the SLS rocket, plus $8 billion in annual operating costs (4 x $2 billion per year), plus a one-seventh share ($4.3 billion) of the development costs, equals $14.3 billion per launch. What was the Mars One budget again? Oh yes. $6 billion. In total. Sorry, but that just won't cut it, will it? Not even close, because NASA reckons it would take at least six of these SLS rockets to deliver all the supplies necessary for a Mars mission. $14.3 multiplied by six equals $73.8 billion. That's twelve times the Mars One budget! And even if, by some miracle of miracles, Mars One had $73.8 billion lying around, the development schedule of the larger Mars-ready SLS means it probably won't be available until the early 2030s. So what else is available in the world of heavy lift launches? Let's take a look at SpaceX.

For those who have followed the world of commercial spaceflight, it will come as no surprise that the SLS is not the only American heavy lift vehicle being developed. SpaceX, whose raison d'être is to get to Mars, is planning to build an entire family of heavy lift launch vehicles that will be powered by Raptor engines. The good news for anyone planning on going to Mars is that SpaceX is planning on having its heavy lift launch vehicles ready in the mid 2020s. As of 2016, the super heavy lift launch vehicle doesn't have an official name, although the space community has dubbed it the BFR, or the Big Falcon Rocket. This BFR, if realized, will be even bigger than the mighty SLS that NASA is planning; a true monster that will utilize nine Raptor engines on a ten meter diameter core, with an option to build a 12.5 meter and 15 meter diameter core. The purpose of such a gargantuan reusable launcher may be to launch the Mars Colonial Transporter (MCT - Figure 7.3).

How will it work? Well, here we enter the world of realities and rumors, but we've heard Elon Musk hinting at developing a BFR capable of ferrying 100 metric tonnes to the surface of Mars. How will SpaceX get that amount of weight to the Red Planet? Well, the fuel will most likely be cryogenic liquid methane, because methane can be manufactured from subsurface ice and carbon dioxide in the atmosphere. Methane also happens to be less challenging to deal with than liquid hydrogen. The engine that will burn this fuel will be the Raptor, various components of which are being tested at NASA's Stennis test stands. The Raptor will likely be available in two versions; a sea-level option that will be used on the first stage and a vacuum option that will be used on the second stage. With these elements in place, Space X can go ahead and head for Mars, and here's how that might play out.

Figure 7.3 Mars Colonial Transporter. Credit: Lazarus Luan

First, the Mars transporter spacecraft will be launched into Earth orbit. This will be followed by the launch of a tanker spacecraft which will rendezvous with the transporter. The second step will be repeated a number of times to fuel up the transporter, after which the transporter will perform a trans-Mars injection burn. As it nears the Red Planet, the transporter will perform a series of entry burns before landing on the surface thanks to supersonic retro-propulsion. After offloading its cargo, the transporter will head back to Mars orbit, perform a trans-Earth injection burn and head home for a landing back on Earth.

Feasible? Perhaps. Don't forget that many in the industry wrote off Elon Musk several times and every time he has come back and proven the naysayers wrong. If there is one thing you don't do, it is to bet against Elon Musk. For the moment, we know a little about the precursor vehicles such as the Falcon X, the details of which have been presented at propulsion conferences over the years. The Falcon X (Figure 7.4) is a two-stage vehicle standing about 75 meters high, with a core diameter of about six meters. It should be capable of lifting 38 metric tonnes into low Earth orbit (LEO). The Falcon X Heavy on the other hand will be capable of launching up to 125 metric tonnes into LEO, and then there's the Falcon XX. Standing 100 meters high with a diameter of 10 meters and powered by six Merlin 2 engines, this monster is advertised as having the power to lift 140 metric tonnes (with three cores the vehicle could lift around 400 metric tonnes!) into LEO. Manned Mars mission material in other words. How much will it cost? Well, we're speculating here, but Musk has been quoted as saying he expects ticket prices to be in the $500,000 range. That would make it affordable for the likes of Mars One but it would put a dent in the spin for the reality television series, because the contestants would no longer be the first to land on Mars. Instead, they would just be tourist class passengers along for the ride with a few other Martian wannabes. Still, nobody is stopping them going off on their own and playing the 'one-way to Mars' game.

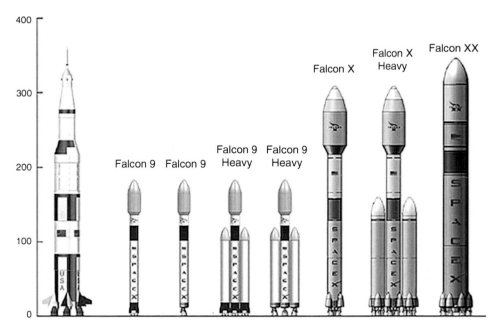

Figure 7.4 Big Falcon Rocket. Credit: SpaceX

SHIELDING

SLS or BFR? Those are the only two viable vehicle options in development for those wanting to fly to Mars. And these puppies won't be ready for a long while. Which is probably a good thing, because there are still plenty of other technological hurdles to be resolved. Let's move on to radiation shielding.

> "Since we're talking here about the risk of developing cancer 25 years or more down the road, there is pretty much zero chance of radiation exposure having any mission impact. Thus, radiation in a Mars mission doesn't really represent a significant danger, as compared with the dangers of launch, landing, and mechanical failure. If you're worried about the risks of a Mars mission, radiation should not be ignored, but it shouldn't be at the top of your list of concerns either."
>
> *Andrew Rader,* Leaving Earth: Why a One-Way to Mars Makes Sense. *2014.*

Priceless. Read that quote again. Carefully. Especially that sentence that states: *'there is pretty much zero chance of radiation exposure having a mission impact'.* Oh Andrew. What have they been teaching you at MIT? I can't be kind about this. Look, there are people who believe humans and dinosaurs roamed the Earth at the same time. They happen to be the same people who watch *The Flintstones* as if it is a documentary. You can cherry-pick the research all you want and come up with some twisted and warped fantasy about how radiation en route to Mars is safe and survivable, but it isn't. Not even close. Read Chapter 6 again if you need reminding. So how to guard against it? Well, we can't. Not yet at any rate, which is why sending a human crew to Mars now would be tantamount to a suicide mission. You can read all the delusional statements you want about how the risk of radiation en route to Mars will induce the same cancer risk as smoking over the same period of time, or that there will enough material on board the vehicle to create a storm shelter, but such a belief is even more of a fantasy than dinosaurs living side-by-side with humans.

Sending a manned mission to Mars today will expose astronauts to lethal doses of radiation. LETHAL. Now you may argue that the premise of Mars One is that the contestants never return, so what if the crew gets fried along the way? I don't have a problem with that, but Mars One is planning on landing at least some of their contestants on the surface. For decades, the radiation problem has been the elephant in the room. And it's a *big* elephant, because there is an awful lot of deadly radiation out there in deep space. Ask any of the world's leading experts if you don't believe me. Ask Professor Kristin Shrader-Frechette at the University of Notre Dame. Go on, call her up or send her an email. Or read the National Academy's Institute of Medicine's report on the problem. Still not convinced? Then send an email to Steve Davison at NASA HQ, who can put you straight on radiation risks. Need to know how dangerous those galactic cosmic rays (GCR) are? Talk to Nathan Schwadron at the University of New Hampshire, or Veronica Bindi at the University of Hawaii. Risk mitigation? Talk to Lisa Simonsen at NASA. Hell, I could go on for pages here. The point is that there are scores of very highly credentialed scientists who have devoted years of their lives – decades in many cases – researching the problem of deep space radiation and they have all come to the same conclusion and it is this: spending six months in deep space will, as sure as eggs are eggs, *kill you*, or at least come close to killing you. But the 'Mars in a decade' crowd just don't listen do they?

"The most challenging medical standard to meet for a Mars mission is that associated with the risk of radiation-induced cancer."

Quote from NASA's Mars Mission
and Space Radiation Risks Overview briefing
to NAC/HEOMD/SMD Joint Committee, April 7, 2015.

Does the 'Mars in a decade' crowd think NASA is making this up? Do they think there is some conspiracy at work here that is manifested by the very best scientists in the world colluding in stating that radiation in deep space is deadly when it isn't? Let's go over this again shall we? The health risks are serious – deadly serious – and while we understand that going to Mars is a risky proposition, these limits are expressed in NASA's 2012 Space Radiation Cancer Risk Model, as recommended by the National Council on Radiation Protection. The calculations used in this risk model take into consideration the full range of solar conditions and shielding configurations for the surface and for the trip to and from the planet. Those calculations have been summarized in Figure 7.5.

As you can see in the diagrams, NASA standards limit the additional risk of cancer death by radiation exposure. But this is not the same as the total lifetime risk of dying from cancer. Let me explain. The baseline lifetime risk of death from cancer is 16 percent for males and 12 percent for females. After a Mars mission, a crewmember's lifetime risk of death from cancer would be about 20 percent. This means that if 100 astronauts were exposed to Mars mission space radiation, between five and seven would die of cancer later in life, attributable to their radiation exposure during the mission, and their life expectancy would be reduced by about 15 years. Now five out of 100 doesn't sound too bad for an

Post Mission Cancer Risk For A 900-day Mars Mission			
Mars Mission Timing	Mission Shielding Configuration	Calculated REID, 95% C.I. (Age=45, Male-Female)	Amount Above 3% Standard
Solar Max	Good shielding like ISS (20 g/cm2) w/no exposure from SPEs	4% - 6%	1% - 3%
Solar Max	Good shielding like ISS (20 g/cm2) w/large SPE	5% - 7%	2% - 4%
Solar Min	Good shielding like ISS (20 g/cm2)	7% - 10%	4% - 7%

Figure 7.5   Post-mission cancer risk. Credit: NASA

undertaking as risky as a manned mission to Mars, but those are the standards NASA abides by. If you happen to inhabit the make-believe world also known as the Mars One organization, you can abide by whatever standards you like, but you'll be taking a risk. Especially if you decide to send females. In the real world of NASA and professional space operations, women qualify for only half of mission assignments because the radiation exposure limit for females is set 20 percent lower than for males. So, if you are an astronaut with two X chromosomes... For some reason, Mars One seems to have overlooked this, or perhaps they just don't care what happens to the female crewmembers. I don't know, because I wasn't there when they concocted this castle in the air.

Oh, but you can protect astronauts, the 'Mars in a decade' crowd whine. Really? How? What robust radiation shielding technologies do we have on the table today? Let's see. Aluminum. Well, that's no good because you need an awful lot of it to provide even basic shielding and it will cost a fortune to send that amount of shielding up into LEO. But what about radioprotective agents? The talk on the Mars blog sites about these magic bullets really underlines just how delusional, deranged and downright uninformed some people can be. There are *no* radioprotective agents that can protect astronauts from radiation during a trip to Mars. Yes, there are all sorts of possible candidates, but none of these have shown any real promise. Phosphorothioates and other aminothiols can be administered before being irradiated, but that's no good if you have no warning of the radiation event. Amifostine, or WR-2721 as it is also known, is no good because like so many thiols, there are all sorts of nasty side effects. Nausea, vomiting, hypotension... shall I go on? Look, if a crew is struck by a radiation event there will be plenty of nausea and vomiting anyway, so no need to make it worse OK? Oh, but we can use antioxidants can't we? Can we? Vitamins A, C and E are routinely touted as being a miracle countermeasure against radiation, together with gluthatione and various phytochemicals and metals (zinc and copper are crowd favorites). True, these dietary supplements have shown some effect against radiation (ascorbate has shown to reduce the frequency of mutations and strawberries have shown to protect the central nervous system against the effects of HZE particles, for example) but there is an awful long way to go before it can be proven than they can protect against cancer risk. The point is that, despite dozens and dozens and *dozens* of research studies on the efficacy of dietary supplements as a means of protecting astronauts against radiation, the most recent findings have revealed that taking these supplements may actually *increase* the risk of death. You see, by rescuing cells that still have genetic mutations, it is possible that genomic instability may occur. Bottom line: all this talk of dietary supplements as a countermeasure for radiation is most likely little more than a therapeutic misadventure. Sure, it makes for great press and a sound-bite saying that strawberries can help protect astronauts, but unfortunately that is what people pick up on nowadays and before you know it, the news is distorted by all those crazy bloggers. And there are some people who actually believe what they read on those blogs. Perhaps the same people who believe in a time when dinosaurs and people roamed the Earth side by side, but I digress.

So what else can be done to protect the Mars One contestants? Well, they could screen for crewmembers who have an increased resistance to space radiation. The problem with this strategy is that there is a risk that the most radiation resistant contestant may also be the least photogenic and that's no good for a reality television show. So no screening then. Which brings us to shielding. The problem with shielding for deep space missions is protecting the crew from GCRs, because high-energy GCR radiation is extremely penetrating. Making matters worse would be the slew of secondary effects caused by the interactions of radiation particles striking the shield. This causes the particles to disintegrate into smaller particles, and while these have lower energy, they can still do some damage. What shielding material to use? That would be hydrogen, which has the highest shielding effectiveness per unit mass. But liquid hydrogen isn't such a great shield material because it happens to be a low temperature liquid, which is why spacecraft engineers prefer polyethylene as a compromise. This stuff is used today in the sleeping quarters on board the ISS. But even this material doesn't stop the high energy GCR, so what to do? Well, the 'Mars in a decade' lemmings will try to convince you that magnetic shielding can be used. Really? It's an idea dreamt up by Wernher von Braun in 1969, but nearly five decades later, where are we with the technology? Not far as it turns out. There are some scientists working at CERN who reckon they might eventually be able to develop a superconducting magnetic shield that will be able to protect astronauts, but turning that concept into reality is still a long way off.

Figure 7.6 Curiosity. Credit: NASA

NOBODY KNOWS HOW TO LAND. EXCEPT SPACEX… PERHAPS

Take a look at Figure 7.6. That's a picture of *Curiosity*. It weighs 899 kilograms and that is the record for the heaviest object landed on Mars. Ever. Not even one tonne! For a manned Mars mission, the minimum mass that must be landed will be at least 15 tonnes. Probably 20 or more. How does Mars One propose to land their contestants? How does it propose to resolve a problem that has defeated the very best engineers and scientists for decades? Let's see what solutions have been put forward by the Mars proponents over the years. Here's one:

> "Upon arrival, the manned craft drops the tether, aero-brakes, and then lands at the 2018 landing site…"
>
> *Excerpt from* 'Human Mars Exploration: The Time Is Now'
> *by Robert Zubrin*

Aero-brakes? Really? Is that the best you can come up with? The author of that nugget of information is Dr. Robert Zubrin, an aerospace engineer and author of *The Case for Mars: The Plan to Settle the Red Planet and Why We Must*. Not by using aero-brakes Bob! You see, aero-braking is just one of many, many, many mission-enabling technologies that have yet to be developed, but before we discuss this, let's remind ourselves of the Challenger and Columbia disasters. These terrible events reminded everyone just how crucial it is for everything – every single thing – to work just perfectly for a spacecraft to travel to space and return to Earth. Whether it happens to be considering the 'what ifs' of an errant block of foam insulation, or predicting the dangers of a damaged seal, the process of sending a spacecraft into space requires that engineers make millions of estimations of what might go wrong during a flight. And despite the best efforts of the brightest and very best engineers on the planet, NASA still suffered two terrible losses. Now let's move on to a manned Mars mission, where those dangers are amplified a hundred-fold thanks to the jaw-dropping complexity of such an endeavor. Let's consider that aero-braking challenge for a moment.

Over the years, we have flown a number of vehicles to Mars and one thing that scientists have learned is that the Martian atmosphere is a big variable – one that is much more dynamic than the atmosphere we have on Earth. Why is the atmosphere so important? Well, most manned mission architectures feature a spacecraft that stays in orbit, loaded with supplies for the trip back home and also to serve as a lifeboat in case things go pear-shaped on the surface. So, to get that spacecraft in orbit around Mars, you have to slow the thing down and it is here that aero-braking could hypothetically be used. Hypothetically. Aero-braking has been used on some robotic missions and this is how it works. The vehicle propulsively inserts itself into orbit before circularizing its orbit. Job done. Sounds simple doesn't it? The problem is that the only way to slow the spacecraft down is to fire retro-rockets, a method that needs an awful lot of fuel which, by the way, has to be lugged *all* the way from Earth! It's a technique that adds weight to a vehicle that is already heavy enough. So how else can you slow the vehicle down? Well, there's aero-capture. This technique uses the drag of the Martian atmosphere to slow the spacecraft. The problem with this is that zipping along at several kilometers per second causes friction, which causes heat, which means you have to lug along a cumbersome aeroshell *and* thermal protection system (TPS) which means more weight! Even if you could solve the aeroshell and TPS problem, how do you devise a software program smart enough to calculate just how deep the spacecraft has to penetrate the dynamic atmosphere and predict when to come back out again to attain the desired orbit? Too deep and you fry the occupants. Too shallow and you don't bleed off enough speed, which means that when you exit the atmosphere you either don't attain the correct orbit or you simply fly by the planet. Now there's a reality show! How do you fine-tune this technique? By sending many, *many* more spacecraft to Mars to characterize the atmosphere. That's one option. Perhaps after 20 or 30 years we will have enough experience to solve the braking problem. Perhaps. But then we're still faced with the EDL problem (Figure 7.7). Again, no word from Mars One on how they plan on fixing this little headache. They had the chance to come clean on this on their website, but there is nothing there. They had another chance when they published their book: *'Mars One: Humanity's Next Great Adventure'*: 277 pages of all sorts of topics about what it will be like to live on Mars, but nothing at all about how the enterprise actually plans to land their contestants on the surface. Like so much in the mission details, Mars One is not firing on all thrusters!

Figure 7.7   An EDL concept for a human landing on Mars. Credit: NASA

"Mars One has developed a realistic plan to establish a permanent settlement on Mars. This plan is built upon existing technologies available from proven suppliers. Mars One is not an aerospace company and will not manufacture mission hardware."

Excerpt from the Mars One website

EDL

So Mars One has developed a realistic plan has it? Really? And where in that plan is the method for landing their contestants on the surface? Where? Existing technologies? Which existing technologies allow us to land a 15-tonne payload on the surface of Mars? Please, enlighten everyone out there. The reality is that there is no existing technology, or combination of technologies, that will allow us to land big payloads on the surface of the Red Planet. But let's indulge the fantasists and the lemmings and imagine that the aero-braking dragon has been killed. We are now in orbit and the contestants have to face the biggest dragon of all: Entry, Descent and Landing, or EDL. Now the 'Mars in a decade' crowd will try to convince you that the vehicle just has to de-orbit, then decelerate using a combination of a lifting entry and parachutes before landing by means of propulsive descent. Simple? In reality, it is anything but. Let's consider some of the constraints. First of all, there is the condition of the Mars One crew. Remember, these contestants have survived a six-month trip in deep space, during which their bodies have been subjected to radiation exposure and prolonged weightlessness. Some may be half blind and some may be suffering from radiation sickness. In this weakened state, the Mars One contestants won't be in any condition to tolerate re-entry decelerations higher than four or five G, and they will only be able to endure this for short durations, even if the loading is through the chest. But when the vehicle transitions from entry to powered descent to landing configuration, it may be necessary to re-orient the crew. This not only causes problems when designing the interior of the vehicle but may actually cause the crew to be disoriented. At such a super critical phase of the flight, this is the last thing a radiation-ravaged and discombobulated crew needs. So how will the Mars One mission planners design a navigable entry corridor that doesn't impose excessive deceleration on the contestants? Well, it's a tough ask.

"Robotic exploration systems engineers are struggling with the challenges of increasing landed mass capability to 1 tonne while improving landed accuracy to 10s of km and landing at a site as high as +2 km MOLA elevation. Subsequent robotic exploration missions under consideration for the 2010 decade may require a doubling of this landed mass capability. To date, no credible Mars EDL architecture has been put forward that can safely place a 2 tonne payload at high elevations on the surface of Mars at close proximity to scientifically interesting terrain."

Concluding remarks from: Mars Exploration Entry, Descent and Landing
Challenges, *by Robert D. Braun, Georgia Institute of Technology, Atlanta,*
and Robert M. Manning, Jet Propulsion Laboratory,
California Institute of Technology, Pasadena, CA

Let's start by imagining the Mars One vehicle bulleting towards the Martian atmosphere about to begin atmospheric entry. During this phase, mission planners must ensure that the contestants are kept safe from thermal loads and that the deceleration forces don't injure or kill the deconditioned crew. The mission planners must also ensure that the vehicle is delivered to an entry point at precisely the correct velocity and angle over the horizon. Too steep and the crew will be fried and knocked out by excessive deceleration. Too shallow and the vehicle will skip off the atmosphere and that will be the end of Mars One. Another important requirement is to ensure all that kinetic energy is bled off so our contestants can look forward to a soft landing. Perhaps Mars One is banking on the Martian atmosphere bleeding off much of that hypersonic speed, but even if the atmosphere can slow the vehicle, there still needs to be some way to steer the spacecraft to a precise landing. When mission planners design a landing on the Red Planet, they refer to the *landing ellipse*. The size of this ellipse is determined by such factors as the navigational capabilities of the vehicle, the weather, and the ability of the vehicle to steer during the EDL phase. A good example that illustrates the concept is the Apollo Command Module (CM), which had to steer during its hypersonic re-entry. Thanks to its navigation system, the CM was guided to a precise landing very close to the recovery ships. Based on the images on the Mars One website, it seems our contestants will also be attempting a landing in a capsule, but how can such a vehicle steer? Well, capsules can 'fly' thanks to reaction control thrusters, which are used to bank left and right. This, combined with retro-propulsion, may be the way Mars One attempts its landing. But before attempting that landing, the vehicle must bleed off all that speed.

Figure 7.8   A disk-gap band parachute. Credit: NASA

Many Mars EDL models use Disk-Gap Band (DGB) parachutes (Figure 7.8) because DGBs tend to perform well at supersonic speeds in a low density atmosphere such as Mars. For an EDL plan that utilizes a DGB, our Mars One contestants would deploy a 30-meter diameter parachute at Mach 3, but even with the deceleration capability of such a system, the capsule will still need an additional means of bleeding off speed. And this is where it gets interesting, because the vehicle doesn't just need to decelerate, it must also perform cross-range maneuvering, which requires acceleration *and* deceleration. It also must be able to perform a search (in hover mode) and ultimately land. Each of these capabilities requires propulsive maneuvering, which requires engines, and these engines must be covered by the heatshield until the entry maneuvers begin. This creates an engineering headache.

Compounding this challenge is the heatshield question: does Mars One use one heatshield (Figure 7.9) for aero-capture and another for the entry maneuvers, or does it use just one heatshield for both? Let's assume Mars One opts for the budget version and only one heatshield is used. The next challenge is fabrication, because that heatshield must be designed to be removed to expose the descent engines once entry maneuvers begin. This is a major engineering challenge and here's why. First, there are two ways to drop the heatshield. One way is to drop it before firing the engines and the other way is to open doors in the heatshield to expose the engines. The problem with the first option is that the ballistic coefficient of the vehicle is greater than the ballistic coefficient of the heatshield and that causes a problem, because until the ballistic coefficient of the heatshield is greater than that of the vehicle, that heatshield will remain forced against the belly of the capsule. Why is this a problem? Well, if the heatshield detaches from the capsule under these conditions, there is a chance the heatshield could ricochet away. So why not go with the door option? The problem with doors is that incorporating these into a heatshield requires adding seams and penetrations and this reduces the effectiveness of the heatshield. But how about positioning the engines on the nose of the spacecraft? Not likely, because this would require rotating the capsule 180° and that would be very disorienting to our already discombobulated Mars One contestants. Of course, all this deliberation about where to position the engines may be a moot point, because we have to consider the mass of the heatshield together with all the other mission components, such as life support, power and fuel. In short, if the heatshield becomes too heavy, then we need a bigger parachute and if we need a bigger parachute, then the chances are Mars One won't be able to land its contestants.

Figure 7.9   The heatshield that will be used on the Orion. Credit: NASA

But let's stay positive and assume Mars One solves the Hypersonic Transition Problem (HTP). It's a stretch I know, but anything is possible. Even a reality television show on Mars. So the spacecraft is zipping along at about Mach 5, the parachutes have been deployed and the vehicle is ready to begin the descent phase. Mach 5 is still a fair rate of knots, and while the atmosphere will kill some of that velocity, the spacecraft will need to use a Terminal Descent Sensor (TDS) to ensure the speed is manageable before landing. This will prove difficult for a vehicle that weighs 20 plus tonnes, because the vehicle will still be moving supersonically very close to the surface. This is known as the Supersonic Transition Problem (STP) and at the core of this challenge is the velocity gap that exists below Mach 5. That's because no supersonic decelerator system exists that can slow a large payload quickly enough. In short, with current EDL technology, a 20 tonne payload bulleting through the Martian atmosphere has just 90 seconds to decelerate from Mach 5 to Mach 1, re-orient itself from a spacecraft to a lander, deploy parachutes to bleed off the speed, search for a landing site, avoid hazards, hover to the landing site *and* touch down. Airbags can't be used because the deceleration forces would approach 20 Gs and most likely kill the discombobulated crew. Parachutes are no good for the deceleration from Mach 5 because they can only be deployed at speeds of around Mach 2.8. A lift vehicle like the Shuttle perhaps? Nice idea, but the vehicle would have to be traveling very, very low to get the benefit of the atmosphere, so by the time the spacecraft had slowed to Mach 2.8 it would be too close to the surface. And even if the vehicle did have time to deploy a parachute, the canopy would need to be 100 plus meters in diameter and there is no known way of deploying a parachute that big. Thrusters then? Would these work? Yes, but the fuel cost would be huge (about six times the mass of the payload!) to ensure landing a 20 tonne payload. But what if Mars One could take along enough fuel? What then? Well, then there is the challenge of dealing with rocket plumes which happen to be extremely unstable, unpredictable and dynamic. Imagine the vehicle streaking nose first through the Martian atmosphere at supersonic speeds with rocket plumes (Figure 7.10) moving around the vehicle. And these chaotic rocket plumes are moving around the spacecraft against extremely high dynamic pressure. The result would likely be loss of the vehicle from the twisting forces imparted by the plumes.

Once again, let's keep imagining and let's fantasize that Mars One can solve the STP as well as the HTP. These guys are on a roll! Now the Mars One capsule is whizzing along at subsonic speed and can switch over to propulsion to adjust its trajectory in preparation for landing. Everything looks peachy, right? Wrong. Even with the STP resolved, the vehicle is just 1000 meters above the surface and that doesn't leave much latitude for maneuvering. As the vehicle slowly sinks lower, our Mars One commander spots a viable landing site, but it is six kilometers away and the wind is kicking up. Getting there is going to eat up an awful lot of fuel, which the vehicle probably doesn't have. But it's not all doom and gloom, because there are a few ideas that may make it possible to get our contestants to the surface in one piece. One such idea is the inflatable supersonic decelerator, also known as the Hypercone. This donut-shaped device would inflate around the vehicle at about 10 kilometers above the surface while the spacecraft is traveling at Mach 4. Acting as an aerodynamic anchor, the Hypercone would slow the vehicle down to Mach 1. Problem solved? Not quite, because this Hypercone would have to be 40 plus meters in diameter and structures that large are a real bitch to control. And even if – a big 'if' – the Hypercone delivers and decelerates the

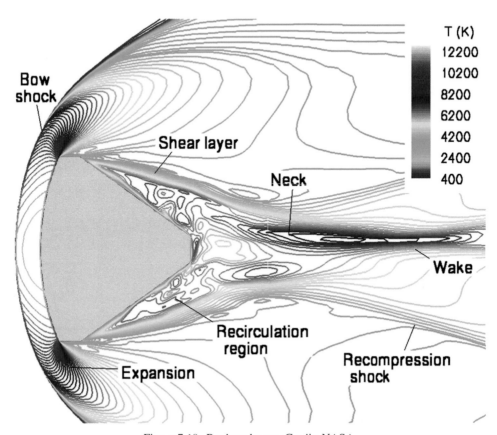

Figure 7.10   Rocket plumes. Credit: NASA

vehicle to sub Mach 1, what then? Parachutes? Not likely, because these take time not only to deploy but also to shed, which won't leave enough time to switch to propulsive systems. And the reason there is so little time is because that spacecraft is plummeting ten times faster than a spacecraft would on Earth for the simple reason that the density of the Red Planet's atmosphere is just one percent of Earth's. But let's continue stretching our imaginations and assume Mars One can solve the entry approach, hypersonic entry *and* parachute descent challenges. Now our contestants are setting up for the landing.

As the vehicle descends to the surface, it will use radiometric data and optical observations to find the landing target. Its sounds very precise, but even with all this information available, chances are the vehicle will be off course once it decelerates below supersonic speed due to the effect of local winds on the parachute. This will be cause for concern, because we know very little about the nature of the winds that blow below 10 kilometers altitude on Mars: it's entirely possible that the vehicle will be blown off course. Way off course. Why is this a problem? Well, Mars One have decided to send the habitats and supplies in advance of their human cargo. If the contestants land 100 kilometers off course, then that's a bad outcome. How long does it take your average Mars One contestant to walk 100 kilometers in a spacesuit after having endured more than six months in space? It's not going to happen. Go back to Chapter 6 and take another look at Chris Hadfield after his stint on orbit. He's sitting in a couch surrounded by medical personnel waiting to whisk him away to the safety of rehabilitation. But there will be no 'whisking away' on Mars for the Mars One contestants will there? Far from it. More than six months of bone loss with resultant high fracture risk and now they have to walk 100 kilometers. Oh dear? Even if they are able to survive the journey without fracture, what about the life support consumables? Say they can make 4 kilometers per hour: that's 25 hours of life support. I don't know of many EVA suits that can carry sufficient consumables to support a trek that long. And this isn't a movie. None of these Mars One contestants will be Mark Watneys. There will be no saving the day. Just a trek to certain death. Great television if you like watching morbid videos I suppose. An interplanetary snuff movie. Groundbreaking, Mars One. Truly groundbreaking.

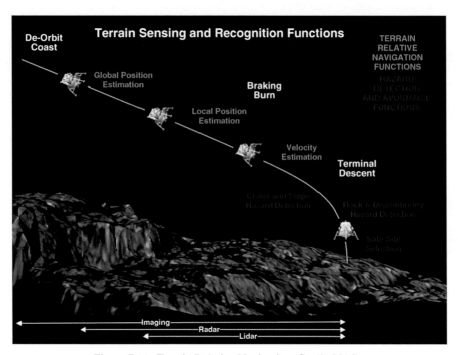

Figure 7.11   Terrain Relative Navigation. Credit: NASA

Now some of you reading this may argue that a lander that weighs 20 tonnes won't be affected by high altitude winds. At least, not as much as a lightweight robotic lander. But how do you know? To what degree has Mars One characterized the upper winds on Mars? Exactly. They know nothing. *Nothing!* But let's not let blind ignorance get in the way of a good story shall we? Although it's probably safe to say the vehicle will be displaced from its target, the Mars One contestants should have been trained in the use of star trackers, altimeters, Terrain Relative Navigation (TRN – Figure 7.11) and velocimeters, so a precision landing isn't completely out of the question. Still, it's a long shot, because the contestants will have to avoid slopes, craters, rocks and all sorts of other natural hazards. Will the vehicle have enough propellant to fly out all the uncertainties as the spacecraft approaches the landing site? How much divert capability will the lander have? One kilometer? Five? Mars One certainly doesn't have a clue because ignorance, apparently, is bliss (see sidebar).

And let's take a closer look at that landing shall we? Remember the first Moon landing. That was flown by one of the greatest test pilots who ever lived. His name was Neil Armstrong. For historians, Neil Armstrong is most famous for being the first man to set foot on the Moon, but in the tight-knit world of test piloting, his exploits as an aviator are still the stuff of legend. Armstrong's capability to memorize the minutest of technical details, combined with his ability to explain the intricate workings of any and every aircraft he tested, made him the outstanding test pilot of his day. Unflappable in the extreme, Armstrong was the very embodiment of the 'right stuff', and the perfect pilot to be chosen as the first to set foot on the Moon. The Mars One crew? How many of them are test pilots? How many are exceptional test pilots? How many of them have spent their careers trouble-shooting problems in high performance aerospace systems? None. Another reason why this little venture is doomed.

Landing technologies

Supersonic Retro-propulsion; Human-scalable EDL architectures; Inflatable Aeroshells; Inflatable Aerodynamic Decelerators; Ballute Technology; Doppler Lidar Descent Sensors; Lander Vision Systems; Conformal Ablator Technologies; Velocity Cancellation And Soft Landing Technology for Safe and Precise Entry Descent and Landing; Lander Vision Systems; Divert Guidance Algorithms for Planetary Pinpoint Landing; Relative Navigation Technologies for Improved Landing Accuracy; Vehicle-to-Vehicle Rendezvous and Docking guidance terminal descent trajectories; etc., etc., etc. The list of technologies required to land humans safely on Mars is a long, long, long one. Even with an Apollo-scale money-no-object approach, it will take decades to bring all these technologies to a level that will ensure even a reasonable chance of landing humans on the surface of Mars. For a commercial venture to iron out all these problems? It's not going to happen. Think I'm exaggerating? Take a look at Virgin Galactic. Remember them? Way back in 2004, Virgin Galactic's Space Ship One rocketed into space, collecting the X-Prize along the way. Not long afterwards, we were promised ticketed rides to space on board SpaceShipTwo – in 2007. Then that became 2009, which became 2010, which... well, you get the picture. Now we are in 2016 and Virgin Galactic is still years away from revenue flights. And they are only going to the relatively easily accessible suborbital part of space. Yet, after hundreds of millions of dollars, four fatalities and with a workforce of more than 500 of some of the most talented engineers in the aerospace arena, Virgin Galactic has still not reached its goal. After 12 years of trying. How on God's green Earth does Mars One hope to get to Mars if Virgin Galactic can't even reach space? This isn't a Harry Potter film: there are no magic wands to wave to solve all those dragons. We can't make them disappear. They are very, very real and they will eat Mars One alive. You see, the Martian devil is in the detail: we do not have the technology to go to Mars and you can warble on about how we do until the proverbial cows come home but it won't alter that irrefutable fact.

"Human space exploration is driven by visions and hopes, but they must be grounded in facts and analysis. Fantasies don't get you to space."

Scott Pace, space policy expert, George Washington University

"Spacesuits: Mars' atmosphere is not suitable for human life. For humans to live and work there, they will need the protection of a full body suit not unlike that worn by the astronauts who walked on the moon during the Apollo program. The Mars Suit must be flexible enough to allow the astronauts to work with both cumbersome construction materials and sophisticated machinery, and at the same time keep them safe from the harsh atmosphere."

Another nugget of information from the Mars One website,
about the need for spacesuits.

LIVING ON MARS

In the spirit of keeping this optimistic, let's presume the Mars One contestants have miraculously survived the landing. Now they have to don their spacesuits and configure the habitats that are waiting for them on the surface. But wait a minute. Do we have a ready-made Mars habitat? Of course we don't. Come to think of it, we don't have any ready-made Mars spacesuits. What planet are these Mars One people living on? The only agency currently developing a spacesuit for use on the Red Planet is NASA, which has come up with the Z-2 spacesuit (Figure 7.12). It doesn't look very futuristic – certainly not nearly as cool as the one worn by Matt Damon in the film *The Martian* – but NASA knows what it's doing when it comes to spacesuit design.

The Z-2 is a super-tough lightweight suit that provides a lot of flexibility and unlike previous spacesuits, this one features an entry hatch which astronauts use to climb into the ensemble. Of course, it's a prototype and it has a long way to go before it becomes operational. For example, you couldn't wear this suit on the surface of Mars today because the current version of the suit has no radiation protective covering. But Mars One won't have access to this suit because NASA is not in the business of selling them. That role tends to be filled by the manufacturers of spacesuits and that happens to be ILC Dover, or Oceaneering International. To give you an idea of how much it costs to design a new spacesuit from scratch, in June 2008 NASA paid Oceaneering International $745 million to create a new spacesuit for the Constellation Program. That was almost ten years ago. Imagine what it would cost now – easily more than a billion dollars, which happens to be a billion dollars that Mars One doesn't have. And even if it did, developing a Mars-ready spacesuit takes a really long time, so no spacesuit for the Mars One contestants.

"Simulation Outpost: Mars One will build several Earth-based simulation outposts for training, technology try-outs and evaluation. The construction technology behind the first simulation outpost will match the simplified level of outpost complexity. The entire outpost will be made of rigid modules – even the 'inflatable' volumes."

Extract from the Mars One website

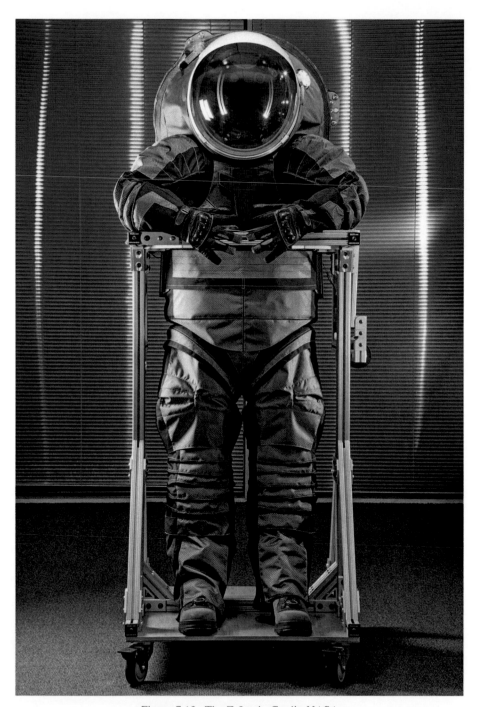

Figure 7.12   The Z-2 suit. Credit: NASA

But let's continue to play along with grand delusion and suppose Mars One manages to re-purpose some old pressure suits that the government had lying around. The contestants are suited up and are making their way to the Mars habitat. But, wait a minute, how on Earth will Mars One have a Mars habitat ready and waiting when no such habitat has been designed? Let me consult the oracle and see what the Mars One book – '*Humanity's Next Great Adventure*' – has to say. Hmmm. 'Technical and Medical Skills', 'Culture, Cohesion and Compatibility', 'With the Whole World Watching', 'Life On Mars'. Where the hell is the section on how you're going to land on Mars and live on the planet? Let's check their website? Nope. Nothing there. Where are your plans guys? Don't have a clue do you? Let me give you the basics. For a habitat, you need to think about structures, command and control, power, life support, accommodation, robotics and automation, EVAs, thermal, and something we in the real-life space industry call In-Situ Resource Utilization, or ISRU. Got that? Good. Now check the following – much abbreviated – shopping list of questions you should being considering:

- What are your subsystem requirements?
- What are your energy requirements and how will you verify these?
- What are your control subsystem interfaces?
- What are your subsystem design issues?
- How will you integrate mission operations?
- What are the crew operations schedules: manned vs. automated?
- What is your schedule template? Presumably this will involve the crew spending lots of time in front of the camera, but things have a tendency to break down in space, so much of their time will be spent on maintenance activities, especially on the life support system, which is another story entirely.
- Characterize all your failure scenarios and address off-nominal conditions
- Does your habitat fit within the dynamic envelope of your launch vehicle? Oh, but you don't have one do you?
- Is your habitat structurally sound in all load environments?
- What is your command and control subsystem?
- Which habitat interfaces will you be using?
- Which user interfaces will you be using?
- What is your plan for your external communications subsystem? This is a television reality show after all, so communication is everything.
- What high gain communications mode will you be using? A Mars orbiting satellite perhaps? Which one?
- And your DSN communication with Earth? How will this be integrated into the Mars One mission architecture?
- Ditto for your local area UHF communications
- How will you supply and transfer power to the habitat(s)?
- Does your supply power have 3-level redundancy? Have you even thought about this?
- Does your power have an emergency power cut-off?
- Which interfaces will be used for local transportation? One rover? two rovers?
- What is your plan to deploy communications hardware and radiator panels?
- How will you externally monitor all your systems and scientific equipment?
- Have you detailed all your worst case thermal scenarios and how these will affect power usage and structural heat losses and gains?

That is a very, *very* short list, and even if Mars One started now it would take the best part of a couple of decades to resolve, even with an unlimited budget. But where are the details? Not in their book. Not on their website. Just vague statements that allude to… well, just the fact that there is no plan, and even if there was a plan, Mars One doesn't have the means to follow through. Look, detailing everything Mars One has to do to have any chance of its boondoggle succeeding would require a major reference work of many thousands of pages and this is a popular science book, so let's just take a look at just one more key technology that is part of the Mars One mission architecture but is still a long, long way from being operational. ISRU. To have an operational ISRU, Mars One *must* have a tried and tested system that can use Martian resources. This system must also be able to provide additional oxygen, nitrogen, and water to the crew in the most austere environment. Let's begin with a question. How many ISRU systems have been tested successfully on Mars? Answer: None. Next question. How many full-scale ISRU systems are planned to be carried on future Mars missions? Answer: Also none. Why? Well, developing an ISRU system is a very tricky business, as NASA and ESA have discovered over the many decades the two agencies have been working on the challenge. The 'Mars in a decade' crowd may think that the ISRU nut is close to being cracked when Space.com posts headlines such as '*Oxygen-generating Mars Rover to Bring Colonization Closer*' (August 1, 2014 by Mike Wall), but 'closer' is a relative term when it comes to ISRU, because 'closer' in this case still means decades. Decades. The next ISRU gadget planned for Mars is a small-scale system slated for launch on board NASA's 2020 Mars Rover. The system, like all systems and just about everything developed at NASA, has its own acronym, MOXIE, which stands for Mars Oxygen In-Situ Resources Utilization Experiment (Figure 7.13).

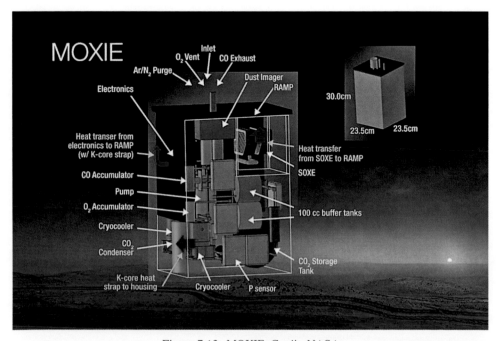

Figure 7.13 MOXIE. Credit: NASA

This little device – and it is very small compared to what a human-rated system will need to be – will process carbon dioxide from the Martian atmosphere and turn it into oxygen and carbon monoxide. All being well, MOXIE will generate about 22 grams of oxygen per hour and will operate for about 7 weeks. Let's put that number in perspective. The average human uses about 550 liters of oxygen per day, or about 23 liters per hour, or about 380 grams per minute. 22 grams per hour? That's less than a thousandth of the amount of oxygen a human would need to breathe. NASA does plan to scale up the technology to a MOXIE that big enough to generate sufficient oxygen for humans. But, as with all these technologies, this will take time. Lots of time.

8

Bursting the Mars One life support bubble

Figure 8.0 Credit: NASA

© Springer International Publishing Switzerland 2017
E. Seedhouse, *Mars One*, Springer Praxis Books, DOI 10.1007/978-3-319-44497-0_8

The Life Support Unit is a Lander rigged with extra technologies which capitalize on the natural resources available on Mars: The *Environmental Control and Life Support System (ECLSS)*. This system uses these natural resources to create a habitable living environment for the astronauts, as follows:

- Electrical energy is generated through the application of thin film solar photovoltaic panels. These are flexible and can be rolled up for compact transportation to Mars.
- Potable water will be created through the heating of water ice in the local ground soil. About 60 kilograms of soil is loaded into a container within the ECLSS by the Rover and heated to evaporate the water. The water is condensed and the dry soil returned to its origin. A portion of the water is stored while a portion is used to produce oxygen. Each ECLSS is able to collect 1500 liters of water and 365 kilograms of oxygen in 365 days time.
- Nitrogen and argon gas are extracted from the Mars atmosphere and injected into the habitable space as inert gases. Remember, 80% of what we breathe on Earth is the element nitrogen.

The Life Support Unit is connected to the Living Unit by a tube which feeds the oxygen, nitrogen, and argon to create a habitable atmosphere. Once the astronauts have landed, it will also be in charge of the water purification and removal of waste gas (carbon dioxide) from the Living Unit atmosphere.

Excerpt from the Mars One website

Sounds so simple doesn't it? But then it usually does on paper. The reality is that life support systems (Figure 8.1) have been a major headache for space missions as long as there have been astronauts. That's because these systems are complicated. Really complicated! Part of the reason for their complexity is the sheer number of functions that must be performed: atmosphere control, supply and revitalization, water recovery and management, temperature control, waste management, food management… the list goes on *and on*. For those of you who aren't familiar with the intricacies of life support systems, the following is a brief overview. Mars One take note, because this will be new information for you!

LIFE SUPPORT 101

Life support systems can be either open-loop or closed-loop. Open-loop systems are those which provide all the resources (water, oxygen and food for instance) and as you can imagine, the amount of resources required increases with mission length and number of crewmembers. Closed-loop systems, on the other hand, only need an initial supply to get the system kick-started. That's because in this system, waste products such as carbon dioxide and urine are recovered and reused. Sounds like Mars One will need a closed-loop system, right? Yes, but the problem with closed-loop systems is that there is no such system with 100 percent closure, which is a system in which *all* the resources are recyclable. In fact, we're not even close to having a system with 100 percent closure, so where does that leave Mars One? Well, it means the Mars One contestants will need to lug along most of their life support consumables (Table 8.1). And that will weigh an awful lot.

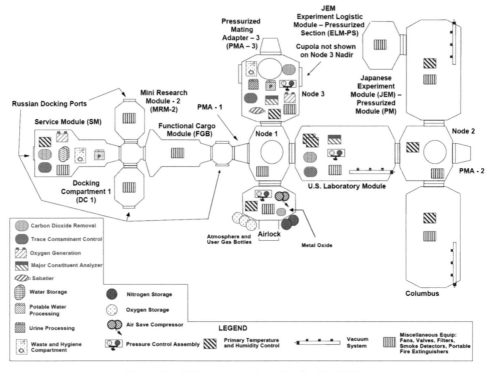

Figure 8.1   Life support schematic. Credit: NASA

Table 8.1   Life Support Mass Balance

Needs	Effluents
Oxygen: 0.84 kg	Carbon Dioxide: 1.00kg
Food Solids: 0.62 kg	Respiration & Perspiration
Water in Food: 1.15kg	Water: 2.28kg
Food Prep Water: 0.76kg	Food Prep Latent Water:
Drink: 1.62 kg	0.036kg
Metabolized Water: 0.35 kg	Urine: 1.50 kg
Hand/Face Wash Water: 4.09 kg	Urine Flush Water: 0.50kg
Shower Water: 2.73 kg	Feces Water: 0.091kg
Urinal Flush: 0.49 kg	Sweat Solids: 0.018kg
Clothes Wash Water: 12:50kg	Urine Solids: 0.059kg
Dish Wash Water: 5.45kg	Feces Solids: 0.032kg
Total: 30.60 kg	Hygiene Water: 12.58kg
	Clothes Wash Water:
	Liquid: 11.9kg
	Latent: 0.60kg
	Total: 30.60kg

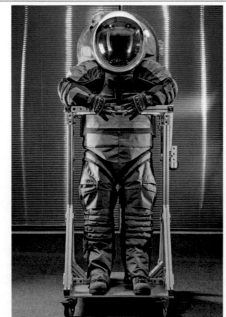

Figure 8.2   Hibernating to Mars. Credit: NASA

The metabolic requirements for a crewmember run something like this: oxygen is about one kilogram a day; food adds up to about half a kilogram; potable water about 2.5 kilograms; hygiene water about 1.5 kilograms. This adds up to a total of 5.5 kilograms per contestant per day, or 22 kilograms per day for the Mars One crew of four. Multiply 22 kilograms by 200 days and you get 4400 kilograms, or more than four tonnes. Now Mars One will want their intrepid contestants to stay alive for at least a few weeks on the surface, so let's give them 10 weeks. Seventy days multiplied by 22 equals 1540 kilograms, which, added to the original 4400 kilograms, brings the total to a whopping 5940 kilograms, or six tonnes. Sorry, but that isn't going to fly, is it? Of course, newer life support technologies are being developed (Figure 8.2), but it is an obscenely challenging business closing in on the Holy Grail of 100 percent closure. Let's put this in perspective for the layman (which includes the Mars One employees judging by the reports of a certain university that studied the Mars One life support plan): if a closed life support system is the summit of Mount Everest, then the current crop of life support engineers are yomping around in the foothills. You may have read popular science articles forecasting the arrival of a closed system in the next few years, but these articles have been written by people who have watched too many science fiction movies and who happen to have the engineering expertise of a seven-year-old and the common sense of a dim six-year-old.

Ever since Project Mercury, spacecraft have used open-loop life support systems. There have been instances when regenerable systems have made the occasional appearance. Molecular sieves for scrubbing carbon dioxide were used on Skylab, amines were used to control carbon dioxide on some long duration Shuttle missions, and more recently, urine

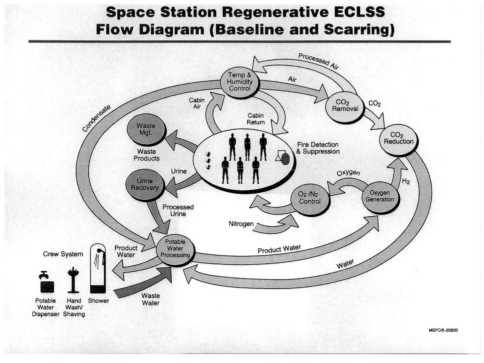

Figure 8.3   Life support on board the International Space Station. Credit: NASA

and waste water has been recycled on board the International Space Station (ISS). But even though the ISS is more of a closed-loop spacecraft than any of its predecessors, the life support systems used today (Figure 8.3) are still mostly open-loop systems. System closure? We're still a long way from achieving that. But before we discuss just how primitive our current life support technology is (compared to those featured in the science fiction movies at least), what exactly does the Life Support System (LSS) have to do?

We'll start with *temperature and humidity control*. This task requires the LSS to remove heat produced by equipment and the latent heat generated by the crew. In today's LSS, heat is removed using heat exchangers, while separating condensed water from the air is the job for the centrifugal separator, which has gained a reputation for being one of the most twitchy of all the LSS devices. Next up is *atmosphere control and supply*. This function is achieved using a system of check valves, relief valves, regulators, sensors and myriad other devices that work – well most of the time – to provide crews with comfortable concentrations of oxygen and nitrogen. Working alongside the atmosphere control and supply system is the Atmosphere Revitalization System, or ARS. Its job is to maintain the quality of the atmosphere, which means keeping carbon dioxide below harmful levels and ensuring that the oxygen concentration of the atmosphere is maintained within a specified range. The ARS (Figure 8.4) also removes particulates and trace gases and is one of the few systems that has a heritage of having had some element that has been regenerable. For example, a regenerable carbon dioxide system featured on board Skylab,

Figure 8.4   Atmosphere Revitalization System. Credit: NASA

a system that was superseded by the molecular sieves that are used on board the ISS. Still, this system has its work cut out to drive down carbon dioxide levels to terrestrial values, as evidenced by the carbon dioxide concentration on board the ISS, which is usually higher than one percent on a good day. Not only that, but the ISS doesn't recover oxygen from carbon dioxide reduction, which is a key process en route to achieving system closure. This means that all the oxygen consumed by the crew, and any animals that happen to be on board, needs to be hauled up into space. This open-loop system results in a horrendous mass penalty.

Trace contaminant removal is next. This system must screen and control materials that might cause harm to the crew. Usually, this is achieved using activated carbon which removes organic contaminants, while chemisorbent beds are used to remove nitrogen compounds and dust particles, and aerosols are removed by screens and High-Efficiency

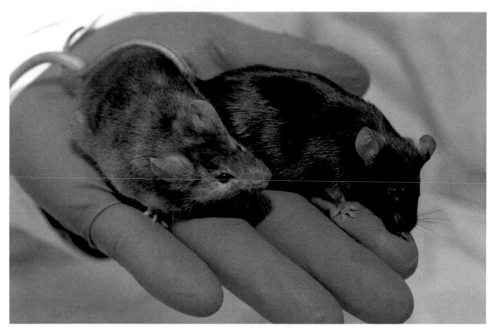

Figure 8.5 Mice: If you are an astronaut on board the International Space Station then you will be drinking recycled mouse urine. Credit: NASA

Particulate Air (HEPA) filters. As you can imagine, this system uses lots and *lots* of expendable materials – stuff the Mars One crew will have to lug along with them inside their spacecraft. Best plan to use a BIG spacecraft! At this stage, there may be some of you who are thinking, 'what about the stories of astronauts recycling their urine: isn't that a closed system?' Well, that would be the *water recovery and management system* (see sidebar). This system bioregeneratively reprocesses urine (human and animal – rats and mice mostly – Figure 8.5), hygiene and washing water which the ISS astronauts drink. Not your taste? Well, then you're not a candidate for a Mars mission. The Mars One contestants meanwhile can look forward to processing each other's urine and bathwater over and over and over… well, you get the picture. In short, yesterday's urine becomes tomorrow's coffee. The system on board the ISS uses a complex process to filter, distill and then oxidize urine. Before this system there was a recycling system that the Russian's used on board their space station *Mir*, but it was always breaking down and didn't turn much urine into water. Not water that any of the crew wanted to drink at any rate!

Now this process of recycling each other's urine may sound like a great way of saving weight, but the system depends on lots of consumables, such as the filters, which cannot be reused. Still, it's a step in the right direction and here's how it works. First the system pulls water out of the urine using forward osmosis, a process that uses salt and sugar to draw the water out of the urine. Enzymes then take over to convert urea into ammonia,

Water recycling system

Until 2003, the crew on board the ISS depended on water ferried to the station via cargo rockets. But in 2003 a water recycling system was delivered, and with this system the astronauts were able to produce 2,700 kilograms of water per year. Today, the ISS has two water filtration systems: one that the Americans use, and one the Russians use. According to Bob Bagdigian, NASA's Environmental Control Life Support System project manager; "The water that we produce meets or exceeds most municipal water product standards." Pretty impressive, when you consider that shower water, sweat, urine and rat urine is reclaimed.

which is then fed into an electrochemical cell that uses that ammonia to generate electricity (one liter of urine filtered in one hour generates a handful of micro coulombs). By all accounts the water tastes fine, and while the system isn't completely closed, it manages to claim about 93 percent of the water on board the orbiting outpost. Meanwhile, scientists are working on other ways to improve the efficiency of water recycling, using processes such as vapor phase catalytic ammonia removal, thermoelectric integrated membrane evaporation and vapor compression distillation. They still have a way to go but one thing is for certain: those traveling to Mars will get there by drinking their own pee.

The next process we'll look at is the *food management system*. All space missions to date have used food that has been produced on Earth. That will have to change, which means the Mars One organization will have to develop some way of producing food inside the spacecraft. Where are we with this technology? Well, don't look to Mars One to do any research into this, but work is underway. Progress is pitifully slow though because there are so many challenges when it comes to growing stuff in space. Part of the reason is the lack of gravity and part of the reason is the spacecraft environment, which must be scrubbed of volatile organic compounds. Then there's the ever-present problem posed by radiation, which can cause mutations and affect growth. One experiment on *Mir* found mutations 20 times higher in space-based tomato seeds than the control seeds stored on the ground. Added these problems are the spectral effects of using only artificial light and not sunlight, and that's just a snapshot of the challenges. In fact, there are so many challenges involved in growing food on the ISS that it wasn't until August 2015 (Figure 8.6) that American astronauts snacked on locally grown product – lettuce in this case – that had been harvested in the station's Veggie plant growth system (Russian astronauts took the risk many years earlier). Cue jokes about 'one giant leaf for mankind'.

The space-borne lettuce was definitely a step in the right direction, but it was a very, *very* small step: lettuce is one thing; growing bigger plants is a whole different ballgame. For a start, you need room to grow crops and that is one thing you don't have on board a Mars spacecraft, where every nook and cranny is at a premium. And then there is the question of environment, because the optimal conditions for growing some plants may be less than optimal for the crew, which means the Mars One enterprise will likely need a separate module. Such a module will also need to be shielded from radiation and will need to provide a plant-friendly environment that will require an awful lot of energy (lighting and

Figure 8.6   Kjell Lindgren and Scott Kelly enjoy a slice of lettuce. Credit: NASA

heat) and water. Work has started, but we are still decades – *decades* – away from achieving full closure, because evolutionary and revolutionary steps are still required in the specialist areas of regenerable processes, multiple redundancy, system reliability, autonomous controls, and integrating the whole kit and caboodle into the spacecraft. Anyone who thinks otherwise has been watching too many science fiction movies and here's why: large-scale space programs take at least a decade from being authorized to first flight, but NASA has no targeted mission after ISS, so there is no clear plan of what to develop and no impetus to develop it. Even if the ISS is granted a life extension to 2028, the chances are there will be no targeted mission for quite a while.

But let's suppose a decision to go to Mars is taken sometime in the 2020s. If that happens, perhaps there will be a chance of the LSS gurus developing the system needed to get to Mars by the late 2040s or early 2050s. Perhaps. That's a bleak prospect if you happen to be Mars One, but the harsh reality is that there are precious few efforts pushing the envelope on closing the LSS loop. And at current funding levels, plans to realize a fully-closed, Mars-ready LSS are overly ambitious with hopelessly unrealistic schedules. Mars One and all the other manned Mars mission advocates can post as many computer-generated images of conceptual life support modules on their websites as they like, but it won't alter the immutable fact that these will *remain* concepts for a long time to come.

"Someone has to ask themselves: Am I ready to rely on this technology which has been tested for two years to operate for an extra 50 years, since my life is dependent on it?"

Sydney Do, MIT

SYDNEY DO AND THE MIT STUDY

Talking of bleak prospects, the proposed life support design for Mars One has come under particular scrutiny and the news is anything but good. The scrutiny came from a group of MIT students, who looked at how Mars One proposed to keep their contestants alive and reckoned the numbers didn't add up. The paper, titled *An Independent Assessment of the Technical Feasibility of the Mars One Mission Plan*,[1] pretty much blew up the tracks in front of the Mars One train, and here's why.

Do's team looked carefully at Mars One's claims that it could establish a sustainable colony using existing technology by the mid 2020s, and they weren't impressed. Since many of the details of the Mars One plan weren't available (they still aren't), Do and his colleagues made a number of assumptions and then proceeded to rip apart the claims that Mars One could support its colonists. First, the claim that "no new major developments or inventions are needed," didn't sit well with Do and his colleagues, who noted that quite the opposite was true: everything from habitation, life support, In-Situ Resource Utilization (ISRU) and space transportation will need to be upgraded. But if current technology was to be used, as Mars One insists it can, Do and his team calculated that 15 Falcon Heavy launches will be needed. This puts a big dent in the Mars One budget. And if by some miracle someone managed to crack the code to growing crops on Mars, the MIT students predicted that this would "produce unsafe oxygen levels in the habitat," resulting in the first crew fatality after less than 10 weeks due to "suffocation from too low an oxygen partial pressure within the environment." And even if by some further miracle the Mars One contestants found some way of surviving (eating each other perhaps?) for, say, 130 months, Do's study calculated that 62 percent of the mission mass will be spare parts. Doom and gloom in other words. Needless to say, Mr Lansdorp was less than impressed, stating that his company didn't have time to respond to all the questions raised by the study and "the lack of time for support from us combined with their limited experience results in incorrect conclusions." That was a pity, because the MIT team actually has an awful lot of experience in campaign analysis for space exploration, with particular expertise in the areas of biomass production, consumption rates, ISRU reliability, and failure analysis. In fact, the MIT team have won awards for their work. Since life support is the key technology to keeping the Mars One contestants alive, let's take a closer look at what the MIT team had to say about how Mars One intends to perform that pretty vital task.

> "In conclusion, this analysis finds that the assumptions made by Mars One do not lead to a feasible mission plan. We suggest modifications to those assumptions that would move the mission plan closer to feasibility. The largest of these is the need for technology development, which will have to focus on improving the reliability of ECLS systems, the TRL of ISRU systems, the capability of Mars in-situ manufac-

[1] Written by Sydney Do, Koki Ho, Samuel Schreiner, Andrew Owens and Olivier de Weck and supported by grants from NASA and the Josephine de Karman Fellowship Trust. The paper was presented at the 65th International Astronautical Congress, Toronto, Canada, in 2014.

turing, and launch costs. Improving these factors will help to dramatically reduce the mass and cost of Mars mission architectures, thus bringing closer the goal of one day sustainably settling the Red Planet."

The MIT's team conclusion on the Mars One
life support plan, such as it is.

One of the main criticisms leveled at the Mars One life support plan by the MIT team was the very low Technology Readiness Level (TRL) of the technology that Mars One proposes to use. For example, Mars One proposes using ISRU to squeeze water from the Martian soil and extract oxygen from the atmosphere. ISRU looks elegant on a PowerPoint slide, but it is a technology that is many, many years from being realized on the industrial scale needed to support a crew of four on a distant planet. Another major criticism is the expectation that the Mars One contestants will live off the land by growing plants. Such a proposal could only be made by someone who has watched way too many science fiction movies, because this is a challenging technology that only exists... well, in science fiction movies. You see – again, Mars One take note – the problem with growing plants is that plants produce an awful lot of oxygen and having too much oxygen in a closed environment is a very, very bad idea, because things have a tendency to spontaneously explode. Now you may think it would be a simple case of venting the excess oxygen, but the technology does not yet exist that can vent the oxygen without simultaneously venting the nitrogen and you need the nitrogen to pressurize the habitats. So, if you use the current mode of venting, the air pressure will eventually become so thin that the Mars One contestants will suffocate (the MIT team calculated this would happen after just 68 days – Figure 8.7). Immolation or suffocation: which way would you prefer to die?

Yet another spanner in the works of the poorly conceived plan devised by Mars One is the idea of sending more contestants: four more every two years. This is madness, because not only will this place ever increasing demands on water and food, but more people means equipment will be subject to more wear and tear, which in turn means more replacement parts must be sent from Earth (Figure 8.8), which in turn means more resupply spacecraft, which in turn means higher mission costs... Oh dear!

The bottom line is that Mars One will be sending four contestants to Mars, knowing full well that their life expectancy will be no longer than 68 days after they land on the surface. Reality television at its rawest. Of course, Mars One won't admit this and Mr Lansdorp was more than a little upset when MIT came out with its paper in Toronto. The whole debacle came to a head on August 13, 2015, at the annual Mars Society Convention, which arranged a public debate between Mr Lansdorp and Sydney Do and Andrew Owens. The subject of the debate? 'Is Mars One Feasible?' The entertainment began with Do and Owens presenting a shortened version of their paper that showed Mars One will fail and fail badly. Not surprisingly, Mr Lansdorp disagreed, insisting that all the technology needed to realize the one-way trip exists, although he failed to offer any specifics, which is something of a theme when it comes to Mars One! Those watching the debate – at least those with any sense – saw the contrast immediately; on one side there were the scientists who had crunched the numbers and on the other was someone playing off the hopes of the 'Mars in a decade' fanatics, who would believe anything you told them.

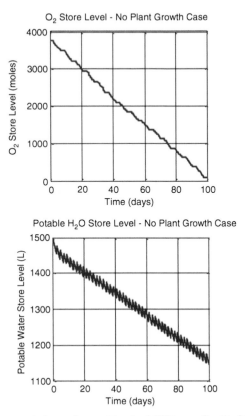

Figure 8.7 The doom and gloom forecast by the MIT team. Credit: Sydney Do/MIT

On the subject of LSS, Mr Lansdorp insisted the current crop of LSS technology is good to go to Mars. Perhaps it is, but the systems on board the ISS break down constantly and need to be repaired, and that can only be done thanks to a huge inventory of spare parts on board the orbiting outpost. And if the astronauts are missing a piece? No problem, because the ISS is resupplied every three months (Figure 8.9). Not so on a Mars mission, which means things will go south very quickly; the MIT team calculated that for the Mars One crew to have even a 50 percent chance of survival, they would need three Dragon capsules full of spares. For every two contestants!

Now back to the ISRU technology mentioned earlier. According to Mars One, the physical processes needed to extract resources from the Martian surface are understood. Well, yes, they are, but the physical processes need to build a warp drive and/or a fusion reactor are understood too and I haven't seen many fusion reactors or warp drives flying around lately. Bottom line: ISRU is at a very low TRL – no systems have been tested in space, never mind on Mars! The same goes for 3D printing by the way, in case you were wondering.

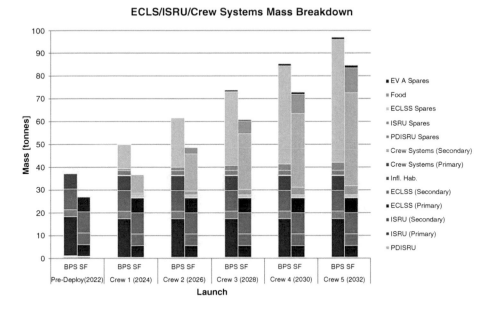

Figure 8.8 More doom and gloom: Mars One just won't work because the spare parts will weigh the mission down. Credit: Sydney Do/MIT

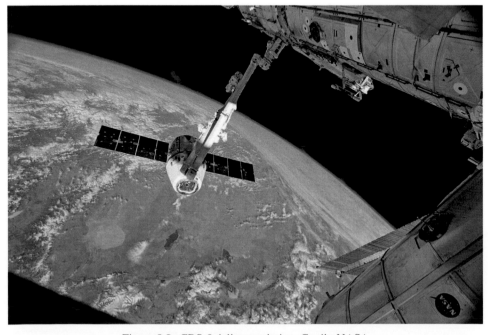

Figure 8.9 CRS-8 delivery mission. Credit: NASA

So why all this BS insistence from Mars One that the technology needed to go to Mars is readily available when clearly it is not? Money. Mars One has none of it and it is trying to spin a good story to try and attract investors. Mr Lansdorp's argument during the Mars Society Convention debate was that once Mars One has money, it will be able to develop the tech. Bit of a Catch-22 if you ask me: no investor in their right mind is going to sink money into a venture that doesn't have a plan and that plan can't exist until the technology is developed. And that will take decades in some instances. Decades!

"My position is that we should get all the technology developed before we go."

Sydney Do, MIT

"These are not impossible [problems] to solve. They are very big challenges that require a lot of researchers and a lot of time. It is our belief that if you're really going to send humans to Mars, you need to go through the technology development and maturation process to enable this."

Andrew Owens, MIT

By now, even the most myopic, scatterbrained dimwit should be able to see the writing on the wall, but I'll spell it out anyway: when it comes to technology, Mars One is putting the cart before the horse. Technology will not somehow miraculously appear when humans land on Mars: that technology must be 100 percent ready and tested *before* humans head off into the unknown.

Epilog

"Flight director: three… two… one… Ignition!

"It's not a real rocket. It's the outside of a real rocket. We did plan to build a real rocket. And that's one thing they can never take away from us. That plan is our legacy.

"Then why did you move up the launch if you knew it was fake?

"To inspire the next generation. And to provide a distraction while we drove away.

"Then why are you still here?

"Our car wouldn't start."

The Simpsons, Season 27, Episode 16: The Marge-ian Chronicles.
Original Air Date: March 13, 2016

© Springer International Publishing Switzerland 2017

E. Seedhouse, *Mars One*, Springer Praxis Books, DOI 10.1007/978-3-319-44497-0

They launched a website, posted a few videos, made vague statements about collaboration with some aerospace companies, listed a couple of resumes of some people with no background in manned spaceflight, and then they staged a press conference. No technical details. No financial disclosure. No plan. But thanks to the media's gullibility and lackadaisical approach to fact-checking, the story is plastered all over the news. The 'Mars in a decade' fanatics are all over it, because no matter how crackpot an idea Mars One is, this group loves fairy tales. Fuel is added to the fire when a Nobel Prize winner says he endorses the half-baked plan and applications flood in by the thousand. From anyone able to operate a keyboard, pay a small fee and upload a couple of documents. Hell, why do you need pilot or an engineer on your way to Mars anyway? Then, after some of the more level-headed journalists start casting a critical eye on the lack of detail and lack of substance of this fantasy, we start reading about concepts like 'redundancy', 'evacuation possibility', 'equipment failure', 'mass estimates', and entry, descent and landing'. Strangely – disturbingly – none of these feature in the Mars One lexicon. Even more worrying is the budget. Six billion. Look, the latest Mars rover cost $2.5 billion, and that was a one-way trip to land a hunk of metal on the surface. Six billion for a human mission? Absurd! And where are they going to get that money? Revenue for a reality television show? Do any of these people have any idea just how boring a six-month cruise through deep space is? How many successful reality television shows do you know that feature contestants pressing buttons and digging holes? But none of this matters to the blinkered 'Mars in a decade' lemmings, whose worldview can't be shaken no matter how outrageous the mission plan – or lack of one in this case. Firmly ensconced inside their fantasy bubble, these lemmings are determined to believe and they spend inordinate amounts of time blogging that determination, as if they are under the illusion that the more blogs they post the more likely the mission will go ahead. But of course it won't. Look at Virgin Galactic as a reference point. It has taken Virgin Galactic more than 12 years to build their spaceship and this puppy is only going to fly suborbital. And they haven't even begun flight testing as of the summer of 2016. As I mentioned earlier in the book, if a manned Mars mission is the summit of Everest, the aerospace companies of the world are still rambling around in the foothills.

We are constantly reminded by the popular science press that we live in a world of miracles but it's really going to take more than one for Mars One to ever get off the ground. In fact, trying to imagine Mars One succeeding brings a new dimension to believing in miracles. That's because the initiative, bold as it may be, is a dream that has no respect for the technology needed to realize it. Pushing the envelope is admirable, but the challenges faced by Mars One are not of the 'don't sweat the small stuff' variety: a manned Mars mission faces hydra-headed problems. Not insuperable, but way out of the league of an organization whose budget is less than NASA uses to put together a PowerPoint presentation. I'm sorry, but it just isn't going to fly, and the reasons why have been discussed in this book. But let's hammer them home one more time anyway. Let's revisit the medical issues.

"When it comes to idiots, America's got more than its fair share. If idiots were energy, it would be a source that would never run out."

The Simpsons Season 27, Episode 16: The Marge-ian Chronicles.
Original Air Date: March 13, 2016

Ever since the US put boots on the Moon, the discussion has been of a manned mission to Mars, but for decades this mission has remained a convenient 20 or more years over the horizon. Why? Human physiology mainly. The human body is an extraordinarily versatile and capable physiological machine, but it was never designed for long duration space-flight. Bones lose strength, muscles waste away, eyeballs become distorted, the immune system takes a hit, and radiation… well, we've talked enough about radiation. Raise this issue with the 'Mars in a decade' crowd and they will point to the wealth of medical data generated by six-month increments on board the International Space Station (ISS), arguing that if astronauts can survive for six months on the ISS, they surely can survive a similar trip length to Mars. But a Mars mission is much, *much* longer than six months. How do we know what happens after six months, or after a year? We don't. Let me put this in some perspective. I used to be an endurance athlete, running 100 kilometer races. I later transitioned to ultra-distance triathlons, racing double ironman, triple ironman and even 10 times ironman races. I can tell you from experience that racing a double ironman is very different from racing two individual ironman races. The point is that it is impossible to extrapolate the six-month data to multi-year missions. Impossible. So all this talk of a one-way mission to Mars is ambition sprinting ahead of common sense.

Of course, saying we can't to travel to Mars in the next few years is blasphemous to the Mars One crowd, many of whom, judging by the online posts, pine for the halcyon days of Apollo. But back in those days, NASA's funding reached 4.4 percent of the federal budget, which equates to about $180 billion a year in today's dollars. And when we did go to the Moon, there was no justification for establishing a colony or even a return visit. But the Mars One evangelists believe we can send humans to Mars, a diabolically more difficult enterprise than landing humans on the Moon, and a much more hostile place to support habitation to boot. Of course this misunderstanding works both ways – the Mars One faithful can't understand the doubters and the doubters have a real problem wrapping their minds around the arguments of the faithful. But Mars One has nothing to do with having an open mind; it has everything do with a reality check.

Now there are some who point out that ten years before astronauts landed on the Moon, NASA wasn't even able to launch a rocket into space, the argument being that given a few more years, Mars One might be able to pull this one-way stunt off. But Mars One won't have the backing of the greatest engineering effort in history that employed 400,000 people and burned up 2.2 percent of the federal budget, so that's that argument blown out of the water. Here's another. With the exception of Harrison Schmitt, every astronaut who walked on the Moon was a stellar test pilot, each of whom was able to process unfathomable amounts of information at warp speed. Mars One's approach? A popularity contest of random Internet contestants. Now I'm not a pessimistic person by nature and I would jump at the chance of traveling to Mars, but NASA spent more than $100 billion in 2010 adjusted dollars just to get to the Moon, and Mars One reckons it can accomplish a Mars mission for what? Six billion? Give me a break! And let's examine the plan for how that money will be raised. The story goes that Bas Lansdorp was inspired to found Mars One after reading about Olympic Games revenues, which were about four billion dollars between 2008 and 2012. If a Mars One mission was to happen, it will almost surely be the most-watched program ever, even surpassing shows featuring the Kardashians. It goes without saying that advertisers will want a piece of that, but how much revenue would such a show generate and how would the show grab the audience's interest?

"This is not about the first footstep and first flag. This is a human mission. These are humans selected by us. You will participate in the experience. That is why it will remain interesting for a long time."

Mars One Messiah, Bas Lansdorp

The way Bas sees it, viewers will be interested in group dynamics and how the crew will behave when the proverbial brown stuff hits the fan. On that score he has a point, because real-life drama is a key element in the success of *Big Brother* and similar reality shows. The problem with Mars One however, is that the human variable is impossible to quantify. And it is partly because of this reason that space-themed reality shows have such a bad history: over the past 15 years there have been almost a dozen such shows bandied about (see sidebar), but not one – NOT ONE – ever saw the light of day. And Mars One is faring little better. Publicity is oxygen for this reality show and that publicity generally needs to be good. But the media are mainly interested in entertainment value, and Mars One has been an easy target for ridicule, which is bad publicity that hasn't helped Mars One.

Spaceflight reality shows

Some of you may remember Mark Burnett`s (producer of *'Survivor'*) idea of *'Destination: Mir'*, which would have flown a contestant to the aging Russian space station in 2001. When a deal couldn't be done with MirCorp, the Russians suggested their own show: *'Ancient Astronaut'*. It also died a quick death. Then, following Mir`s demise, Burnett revised his idea with *'Destination: Space'*, which would have flown a contestant to the ISS. Hot on the heels of Burnett`s venture was *'Celebrity Mission'*, which would have placed Lance Bass in orbit. That failed too. As did *'Astromom'*, an effort to send Lori Garver into space. A European idea called *'Space Commander'* was quickly forgotten, but then came the success of SpaceShipOne, and many naïve people reckoned there would be reality shows tied to suborbital space operators. They were so wrong, because 12 years later those operators are still years away from revenue flights. That didn`t stop Mark Burnett though, who pitched a show called *'Space Race'* that was rumored to have been picked up by NBC. It wasn't. Nor was *'Milky Way Mission'*, a Sony Pictures television project that would fly contestants on board the Lynx.

So, space-themed reality shows have been spectacularly unsuccessful, but Bas is insistent Mars One can be different. But how? Well, the visuals could be entertaining. You could show the candidates being put through their paces during survival training: eating bugs, setting traps to catch rabbits, and huddling naked in pairs inside sleeping bags while trying to keep warm during arctic survival come to mind. You could put them in a centrifuge and show them undergoing really unpleasant medical tests, such as enemas and appendectomies. You could strap them into high performance combat aircraft to expose them to unusual attitudes and various vomit-inducing positive and negative Gs. You could show them flying parabolic flights to see if they can hold on to their breakfast. You could... well, you get the idea. The problem with all these tests – all staples in any astronaut training program by the way – is that they cost money, and one of the reasons reality television exists is because it is cheap to produce. Nothing about astronaut training is cheap. Nothing at all! And training will only be interesting for a short while. What makes reality shows successful are fights and drama – that's what gets the ratings and generates advertising revenue, which in turn result in profits. But you don't need to work for NASA to know that the very last thing you need on a trip to Mars is a fight.

And then there is the schedule. What schedule? When is it going to start? In 2020? 2025? Who knows? And when – if – it does finally get going, how will Mars One sustain interest for the ten years preceding the mission? Ten years! And then there's the money. Bas holds up the Olympics as his revenue model, but to say that your reality television show can make money simply because the Olympics does is akin to saying your school soccer team can make money televising its games because the FIFA World Cup makes millions. Bottom line: Mars One and the Olympics are not in the same league. Not even close. Investors? No, that won't work and here's why: what investor wants its product associated with an enterprise that ends up killing people? Don't forget that the demise of these contestants will in all likelihood be very public and very graphic: punch in 'radiation sickness + images' into Google in case you don't believe me. You see, when things go pear-shaped in 'Survivor', a ship comes along and the contestants are rescued. That will not be the case for Mars One.

> "Barry: The hab study is complete. To all the male participants, your monumental incompetence has sullied and cheapened space forever. Now get out. Good. It's a stupid idea and I hope everyone dies."
>
> *The Simpsons Season 27, Episode 16:* The Marge-ian Chronicles.
> *Original Air Date: March 13, 2016*

So, in light of all these negatives, why is there still so much interest in a venture that has about as much chance of succeeding as a drunken wildebeest wandering into a bar-beque afternoon for a pride of lions? I have my theories. One theory is that there has been so much talk of manned missions to Mars for so many, *many* years, that there is now a sense of desperation for somebody – anybody – to land on Mars. Suicide mission, bare bones mission – it doesn't matter: just get us there already! And I think the Mars One 100 candidates realize this, which is why they continue to sing from the one-way mission hymn book. Look, in most cases, these Red Planet recruits are smart enough to know that Mars One hasn't a snowball in hell's chance of succeeding, but they want their 15 minutes of fame. They know they'll never get that by being a real astronaut, because very few of them have the qualifications for that job, so Mars One provides the perfect vehicle for fleeting celebrity and they seem determined to milk it for all its worth.

For this group, it doesn't matter how outrageous the enterprise is, or how absurd the Mars One announcements are about the timeline: this group wants to ride the fame train as long as possible. Never mind that Mars One has yet to figure out the design of the habitat, the landing module, or any other item of hardware for that matter. For this group, putting the cart before the horse seems to be the way to do business, which is why all their efforts are spent on highlighting the science fiction details of the mission plan (pick up a copy of the Mars One book and read it for yourself). Any sane person with even the flimsiest grasp of spaceflight basics will realize that no-one goes from zero to Mars with nothing in between. Just take a look at SpaceX, which is probably the likely candidate for landing humans on Mars first. For years, Elon Musk has been steadily developing the systems and hardware that will be used to ferry humans to the Red Planet: this is real world spaceflight. Meanwhile, all Mars One has to offer is the occasional statement from Bas about how a Mars mission will inspire students to believe that anything is possible and that... yawn.

So where does all this leave Mars One? Well, there is always the chance it could get off the ground as a half-assed, half-baked, half thought-out mission. This will be good news for the fame whores, who will get their 15 minutes of notoriety before dying lingering and agonizing deaths (all on prime-time mind you!) somewhere en route to Mars (check out those images of radiation sickness again). Perhaps Bas could join them. I know a lot of people who would pay money to see that.

"There's a great self-defeating optimism in the way this project has been set up. I fear it's going to be a little disillusioning for people because it's presented as if it's going to happen and so all those people are excited."

International Space Station Commander, Chris Hadfield

But while Mars One is surely destined to fail, there are some who argue that the venture can succeed even in failure. How? Well, the argument goes that by announcing their plans in such a public forum, it can be said that Mars One will have done its bit in chipping away at the business of manned planetary exploration, which has historically been the preserve of governments and space agencies. Continuing this argument, a case can be made that the audacity of Mars One may help inspire public enthusiasm for an eventual manned mission to the Red Planet, perhaps making it a little easier for the next group that comes along with a plan. The counter to this argument is the likelihood of Mars One failing, as it surely will. When the Apollo 13 mission suffered a near mission-ending event, NASA didn't close down mission control because the television ratings were down. They helped the astronauts every step of the way back. Mars One? When the crew starts dying of radiation sickness, it will be broadcast on prime-time, and some of those watching will be the bean-counters who ultimately approve funding for manned missions. How do you think having those images broadcast on prime-time, and the ensuing public trauma, will affect the likelihood of *any* future missions being given the green light?

Look, even the most upbeat optimist would find it almost impossible to conceive of an outcome that isn't tragic when it comes to the subject of Mars One. Yes, the hardest part of a manned Mars trip is the return, but nixing the return doesn't mean all your troubles evaporate. And even if they did, how is it possible not to be suspicious of an organization that has no rockets or spacesuits, or much of anything, but is insistent it will be sending humans to Mars? It's all too good to be true, especially now that contracts with Lockheed Martin and *Big Brother* producers Endemol have fallen through. Mars One is a lot of hot air. At best, a thought experiment that has been the catalyst for reopening the conversation about sending humans to Mars: at worst, an elaborate hoax about sending four kooky misfits on a suicide mission.

Appendix A

Mars One reveals new details on Astronaut Selection Round Three

This appendix includes features about astronaut selection and training that appear on the Mars One website. My comments on these processes appear in a series of sidebars.

Amersfoort, June 6, 2016: Mars One released new information about the third round in the Astronaut Selection Program during a private Mars One event in Amsterdam. The third selection round is designed to trim down the remaining 100 candidates to forty through a series of group challenges. The candidates will compose the groups for the third round themselves.

Over the course of five days, candidates will face various challenges. It will be the first time all candidates will meet in person and demonstrate their capabilities as a team. Candidates will start the group challenges in 10 groups of 10. These groups will change throughout Round Three due to continuous elimination, and the selection round will end with 40 candidates.

In this round the candidates will play an active role in decision making/group formation. Mars One has asked the candidates to group themselves into teams with the people they believe they can work well with. All groups have to adhere to certain criteria, such as a gender ratio of 50/50, as well as maintaining age and nationality diversity. The self-selection placement has already started.

> "We want the groups to be as diverse as possible, and to utilize the uniqueness and special contribution from, for example, different backgrounds in order to solve complex problems, as a continuation of the work in JAXA and NASA."
>
> *Mars One's Chief Medical Officer, Dr. Norbert Kraft.*

The majority of the challenges Mars One plans to conduct were previously used in a study by NASA in order to determine:

- The best crew/crew combination
- The best selection tools
- The best training method for long duration space flights.

© Springer International Publishing Switzerland 2017

E. Seedhouse, *Mars One*, Springer Praxis Books, DOI 10.1007/978-3-319-44497-0

Indoor and outdoor group challenges will, amongst other things, test the candidates' ability to work in a team within limited conditions, interdependency, trust, their problem-solving and creativity skills, their thoroughness and precision, and their clarity and relevance of communication. The candidates' knowledge of provided study materials is essential to progress in the challenges. Candidates are eliminated based on their behavior both inside and outside the group challenges, which will be reviewed by the selection committee. At the end of each day, a sociogram will be used to explore the candidates' preferences for whom they would like to work and live with, and this will be taken into consideration by the selection committee when deciding whom to select out. Every day, ten to twenty candidates will leave the selection.

The selection procedure will provide insights into group dynamics. How did the candidates organize themselves into teams? How well did they solve problems as a team? How did each candidate handle the conflicts that inevitably emerge when facing a challenge together?

The Mars One candidates come from all over the planet, and will undertake the long journey to Mars to live there for the rest of their lives. Mars One selection committee members Norbert Kraft, M.D., Prof. Raye Kass, PhD, and James Kass, PhD possess understanding of different cultures as well as many years of experience working with extreme environments, and, of utmost importance, isolated habitats. They have professional experience in the field of Human Space Flight (Group Dynamics/ Long duration Space Flight/ Medicine/ Psychology/ Psycho-physiology) and extensive work with astronauts from JAXA, NASA, CSA, ESA, and RFSA.

From this selection round onward, the selection procedure and training activities of the astronaut candidates will be filmed for audiences across the globe. [The] 40 remaining Round Four Candidates will begin the isolation portion of the screening process. The results of the isolation challenge will reduce the 40 candidates down to 30 who will then undergo the Mars Settler Suitability Interview.

Isolation

Isolation analogs are nothing new and are being used by NASA today to train astronauts for long duration missions (NEEMO for example). But the Mars One isolation is measured in days, not months: to provide added realism – and drama (!) for those watching this on TV – this isolation challenge should be a real challenge in the order of six months (minimum) in a hostile environment such as the Antarctic.

Choosing the Mars Four

Here Mars One Chief Medical Officer Norbert Kraft, MD, explains the training process and selection of the final four candidates for travel to Mars.

"We will end up with 24 candidates out of the original 200,000-plus applicants. Some of the details concerning training and competition still need to be discussed with the organization that gets the broadcasting rights, because some types of challenges and activities are more telegenic than others. However, we know what they must learn. The basic training will involve the Mars 24 living and studying together in a Mars One facility where their families can join them. Each year, the 24 will study the skills and knowledge that they need to be self-sufficient on Mars — medicine, dentistry, agronomy, electronics, political science, law, and so forth. They will train together for nine months each year. Since team members will come from all over the world, we will be sure that some of the training occurs in the home culture of each team member. For example, if a candidate is from Russia, that candidate's team will spend some time in Russia, so they can become familiar with the culture.

"Three months a year will be spent competing with other teams, and this will be tricky. Six teams of four members will compete to win at challenges related to what they have studied over the previous nine months. The teams will compete against each other, but there will also be times when two of the teams have to collaborate. This is important, because we expect to send new teams of settlers to Mars every four years. This means it is critical that they are good at figuring out how to welcome new team members and collaborate after they've developed team cohesion.

"When they compete, we will be studying their team dynamics and their success at the challenge they were given. We are interested in how they succeed, but also what they do about problems they can't solve. How do they pull themselves together? How do they change their team approach when faced with problems they can't solve? Each group selects its own approach — anarchy, military, democracy can all work — but they have to perform better than the other teams. So the capacity to change organizational style to fit the challenge is critical, and it is this capacity to adapt that will serve them best in the harsh conditions on Mars. They will learn this capacity through the challenges, and because they learn this through experience, it will stay with them.

"Each team will have a specific trainer to work with them, and the teams will compete to continue training for the Mars mission. This means some teams will not continue after the competition, and new teams will be coming on as Mars One opens the application process for new candidates.

"When it comes to judging the success of the teams, we envision five votes to select the winning teams each year. One vote comes from winning the competition. Three votes come from judges with special expertise in the areas being challenged. And the public gets the fifth vote. The public can be the tiebreaker."

Selection

It's not the way governments select their astronauts, although crew resource management is a key element in selection. But where will these exercises take place? The 'harsh conditions' of Mars is mentioned, so I would hope these team dynamics exercises will play out in a hostile environment as well. And giving the public the 'tiebreaker' vote? You just know that the one best qualified to progress will be ditched in favor of the one that best plays the crowd, because that's what reality TV audiences do.

What are the qualifications to apply?

Qualifications

Mars One will conduct a global search to find the best candidates for the first human mission to Mars. The combined skill set of each astronaut team member must cover a very wide range of disciplines. The astronauts must be intelligent, creative, psychologically stable and physically healthy. [Here] Mars One offers a brief introduction to the basics of our astronaut selection process.

The astronaut selection process

In spaceflight missions, the primary personal attributes of a successful astronaut are emotional and psychological stability, supported by personal drive and motivation. This is the foundation upon which a mission must be built, where human lives are at risk with each flight. Once on Mars, there are no means to return to Earth. Mars is home. A grounded, deep sense of purpose will help each astronaut maintain his or her psychological stability and focus as they work together toward a shared and better future. Mars One cannot stress enough the importance of an applicant's capacity for self-reflection. Without this essential foundation, the five key characteristics listed (see table A.1) cannot be utilized to the fullest potential.

Table A.1 Five Key Characteristics of an Astronaut

Characteristic	Practical Applications
Resiliency	Your thought processes are persistent.
	You persevere and remain productive.
	You see the connection between your internal and external self.
	You are at your best when things are at their worst.
	You have indomitable spirit.
	You understand the purpose of actions may not be clear in the moment, but there is good reason—you trust those who guide you.
	You have a "Can do!" attitude.
Adaptability	You adapt to situations and individuals, while taking into account the context of the situation.
	You know your boundaries, and how/when to extend them.
	You are open and tolerant of ideas and approaches different from your own.
	You draw from the unique nature of individual cultural backgrounds.
Curiosity	You ask questions to understand, not to simply get answers.
	You are transferring knowledge to others, not simply showcasing what you know or what others do not.
Ability to Trust	You trust in yourself and maintain trust in others.
	Your trust is built upon good judgment.
	You have self-informed trust.
	Your reflection on previous experiences helps to inform the exchange of trust.

(continued)

Table A.1 (continued)

Characteristic	Practical Applications
Creativity / Resourcefulness	You are flexible in how an issue / problem / situation is approached.
	You are not constrained by the way you were initially taught when seeking solutions.
	Your humor is a creative resource, used appropriately as an emerging contextual response.
	You have a good sense of play and spirit of playfulness.
	You are aware of different forms of creativity.

Characteristics

Yes, these are all good characteristics but a Mars crewmember will require other attributes: high resistance to radiation, high bone density, and being male (females will not be going – even those with the flimsiest knowledge of space life sciences know why)

Age

The astronaut selection program will be open for applicants who are 18 years or older. This is the age by which children become legal adults in most countries in the world. Mars One believes it is important that applicants who enter the astronaut selection program are capable of entering into a legal contract without the supervision of others.

Medical and Physical Requirements

In general, normal medical and physiological health standards will be used. These standards are derived from evidence-based medicine, verified from clinical studies.

- The applicant must be free from any disease, any dependency on drugs, alcohol or tobacco
- Normal range of motion and functionality in all joints
- Visual acuity in both eyes of 100% (20/20) either uncorrected or corrected with lenses or contact lenses
- Free from any psychiatric disorders
- It is important to be healthy, with an age- and gender-adequate fitness level
- Blood pressure should not exceed 140/90 measured in a sitting position
- The standing height must be between 157 and 190 cm.

How will the astronaut selection proceed?

The selection process is made up of four rounds.

1. During the initial round all candidates must submit an online application that consists of general information about the applicant, a motivation letter, a resume and a one-minute video that answers provided questions and explains their reasoning.

2. Candidates making it to the second round are required to obtain a medical statement of good health from their physician and will be invited for an individual video interview with Mars One.
3. During the third round candidates will participate in group challenges and will be par-take in an in-depth interview.
4. The final selection creates international groups of four candidates who are expected to demonstrate their ability to work together and live in harsh living conditions. The teams will also receive their first short term training in a Mars outpost.

Selecting the Crew

There are multiple the requirements to become a Mars One astronaut. Applicants' characteristics must fit with those of an astronaut. Meaning the candidate needs to be:

- Resilient
- Adaptable
- Curious
- Trustworthy and Trusting
- Creative/Resourceful
- Above the age of 18
- A2 English level

Selection Process: from Round Three onwards

Round 3

The third round is an international selection round. Candidates who make it into this third selection round will participate in group challenges that demonstrate their suitability to become one of the first humans on Mars, and will take part in longer and more thorough interviews. The Mars One selection committee will determine who will pass to the final selection round.

Round 4

The Mars One selection committee will create international groups of four candidates. The groups will be expected to demonstrate their ability to live in harsh living conditions, and work together under difficult circumstances. The groups will receive their first short term training in a copy of the Mars outpost.

From the first selection series, up to six groups of four will become full time employees of the Mars One astronaut corps, after which they will train for the mission. Whole teams and individuals might be selected out during training if they prove unsuitable for the mission.

How are the astronauts prepared?

Mars One Astronaut Training Program

- After the crews are selected they begin training. The training consists of three (phases) including technical, personal, and group training.
- Phase 1: Technical training includes the training of two crewmembers to be proficient in the use and repair of all equipment to the extent that they can identify and solve technical problems. In addition, two crewmembers will receive extensive medical training in order to treat minor, major, and critical health problems. At least one person will train in the studies of Mars geology and the remaining person will gain expertise in exobiology, which is the biology of alien life.
- Phase 2: Personal training consists of ensuring that the astronauts are able to cope with the difficult living environment on Mars. Since these individuals will be unable to speak to friends and family on Earth face-to-face, a certain amount of coping skills are essential.
- Phase 3: Group training will mainly take place through simulation missions. During these simulations, the astronauts take part in a fully immersive exercise that prepares them for the real mission to Mars. The simulated environment will invoke as many of the Mars conditions as possible. Immediately after selection, the groups will participate in these simulations for a few months per year.

Phase 3

Phase 3 promises to be interesting, especially the section that promises to invoke as many of the Mars conditions as possible. Presumably (hopefully) this will mean our intrepid reality show contestants will be performing in a lethal radiation environment where the barometric pressure is maintained at one percent of Earth's atmosphere. After all, realism is everything in a reality TV show, isn't it?

Appendix B

Selected profiles from the Mars One 100

To give readers an insight into the quality of candidate that made it into the final 100, what follows is a snapshot of ten of the Mars 100 candidates. Biographies listed are taken from the Mars One Community Platform: https://community.mars-one.com/

Profile 1: Shirelle.

Shirelle seems to be keen and eager to rack up the qualifications needed to make her one of the final candidates, although her grammar skills seem to be lacking (tense Shirelle!). Still, which one of the current crop of reality show contestants has good grammar skills?

Name:	Shirelle Erin Webb
Sex:	Female
Age:	23
From:	United States
Language:	English

Self Introduction:

I know that there are very few individuals that understand why we want to go to Mars. There are very few individuals who understand why anyone would put money into this program. The problem is that our generation has fallen out of love with knowledge and adventure. People forgot that all our technology and advancement came from those who desired more.

Being involved with Mars One is an opportunity to re-inspire the next generations and hopefully have them learn more. This is a chance to learn more and go farther than we could ever imagine.

© Springer International Publishing Switzerland 2017
E. Seedhouse, *Mars One*, Springer Praxis Books, DOI 10.1007/978-3-319-44497-0

I am trying to do everything I can to be a vital part of the Mars One team. I have completed my paramedic courses and am about to take my certification tests. I have worked with the Corpus Christi Fire Department as a first responder while doing my internship. I am pursuing a degree in pre-med to compliment my studies in paramedic medicine. I will switch to post graduate studies in physics which should give me a thorough foundation in medicine, science, and environmental research.

Interests:

I enjoy reading science fiction stories. My favorite book is "John Carter of Mars." I enjoy super-hero comic books and I love to write stories. I enjoy running and yoga.

Profile 2: Maggie.

Maggie has all the qualifications for… well, for making costumes I guess, judging by her resume, although her engineering skills should come in handy when the life support systems start breaking down, as they surely will.

Name:	Maggie
Sex:	Female
Age:	31
From:	United States
Language:	English

Self Introduction:

Raised on a farm in the heart of the US, my first word was 'home' while pointing up at the stars. Although I graduated Cum Laude with a degree in Electrical Engineering, I put my technical skills into making high-end costumes, starting an international business selling the largest zippers in the world. My passion is adventure and my strengths are optimism, intelligence, and creativity… which help me transcribe my journey into art through drawing, story and song.

Interests:

Physics, chemistry, math, technology, music/frequency analysis, electromagnetics, robotics, drawing, writing, acting, singing, percussion, clothing/costume design and construction, cooking, camping, traveling, kendo. Transformers, Star Wars, Star Trek, Mass Effect, Tolkein.
   mars-maggie.tumblr.com

Profile 3: Zachary.

Zach has one of the most Mars-specific resumes, although he is light on qualifications, which isn't surprising since he is only 29. While on paper having a geochemist on the team may sound like a good idea, the first few teams landing won't be doing much else apart from fixing life support equipment… and simply surviving.

Name:	Zachary
Sex:	Male
Age:	29
From:	United States
Language:	English

Self Introduction:

I have an unmatched passion for exploring the universe. Solving complex problems, both individually and as a member of a team, is what I do best. I am very fit and can withstand grueling physical activity. My temperament is calm and relaxed, handling the most stressful situations with ease. I am a very clean and organized person, important habits for long duration space missions. The background I bring from planetary science will be invaluable to the Mars One mission; I am a geochemist and field geologist specializing in Mars and Moon studies. Currently, I am at the University of New Mexico working on Mars Science Laboratory rover operations for the ChemCam laser instrument. Other projects include landing site analysis for future manned and unmanned missions to Mars, analyzing terrestrial impact-induced hydrothermal processes/mineral assemblages, and conducting simulated rover and manned missions at terrestrial analogue sites.

I have been training for this mission my entire life, and I am ready to go to Mars.

Interests:

My interests on Earth include: bicycling, running, scuba, golf, horticulture, playing guitar (https://manonmars1.bandcamp.com), rock climbing, studying geology, collecting minerals, and exploring the wilderness.

My interests in space include: astrogeology, astrobiology, in-situ resource utilization, giant impact processes, terrestrial analogue comparison, and planetary mission operations.

Profile 4: Divashen.

Yes we could begin a new age of astronomy and science Divashen, but you and your Mars One crewmates won't be playing a part of that because you will be on the very threshold of survival. Playing guitar and video games are all well and good, but don't add up to a bag of beans when the carbon dioxide scrubber has gone down for the umpteenth time and poor old Lucy is dying of radiation sickness and spilling her guts on the floor for the third time that morning.

Name:	Divashen
Sex:	Male
Age:	24
From:	South Africa
Language:	English

Self Introduction:

My name is Divashen. I was born on the 23rd of January 1992 in Westville, Durban which is situated on the beautiful sunny east coast of South Africa. I [am] studying a Bsc degree double majoring in applied mathematics and physics at the University of KwaZulu-Natal. My intention is to go into astropyhsics. I love challenging myself and having new experiences. Space and space exploration has always been a dream of mine. To set foot on another planet and pave the way for mankind's journey into a true space age would be so surreal. I believe that we can responsibly inhabit other planets and advance our knowledge of science along the way. On Mars we would be able to begin a whole new age of astronomy and science.

Contact me via: Facebook: Divashen Govender
Email: divashen.govender@gmail.com
Twitter: @DivashenG

Interests:

I have a keen interest in astronomy, reading, playing guitar, surfing, video games and relaxing with friends. But my obsession and biggest interest is space and what lies beyond. The thought of going beyond our world has been a dream of mine for as long as I can remember. That is my ultimate interest.

Profile 5: Gunnar.

Not much in the qualifications department for Gunnar, although his penchant for road-trips will no doubt stand him in good stead – after all, Mars One is first and foremost a reality show, and second, a road trip… with a difference. Again, this talk of inspiration is all well and good, but I'm not sure how many will be inspired by crewmembers projectile vomiting and dying from asphyxiation. Still, it takes all sorts I suppose.

Name:	Gunnar
Sex:	Male
Age:	43
From:	Australia
Language:	English

Self Introduction:

The day I was born I was out exploring the world, society and nature with all their beauty always looking for solutions and new adventures. I never stopped asking questions and I never lost my curiosity. Independent and free minded I try to see life from a different point of view. I was born in East Germany and migrated to Australia in 2003.

When I first heard about Mars One I was hooked and after I read the qualifications to apply it felt like it was written for me. Imagine Mars One is successful and we're going to build the first human settlement on another planet. Wouldn't it be wonderful if we would create a second home with new ideas and new ways of thinking? We have the whole human history we can learn from and our world needs a positive event much bigger than the moon landing and all the sports events put together very soon to inspire every one of us to change our thinking and our way of living because the only hope for human kind lies in the transformation of the individual. Could Mars One be this event?

Interests:

I'm a big fan of road trips, nature, the universe, hiking/bushwalking, little bit of slacklining and photography, critical thinking, science, philosophy, travelling, documentaries, a good video game...

Profile 6: Yari.

Another engineer. Great. Mars One will need a bunch of them because of the aforementioned life support challenges. Working with decision support tools is an advantage because this mission will run on them. Keeping a healthy body will be difficult though Yari, so I suggest you read a few chapters about what radiation does to human physiology: be prepared because it ain't pretty!

Name:	Yari
Sex:	Female
Age:	28
From:	United States
Language:	English

Self Introduction:

Hello! I am future Mars settler Yari. I hold a bachelor's degree in Engineering Science from Smith College. By day I am a data analyst working for air traffic control decision support tools and analyzing the environmental impacts of aviation, by night I am a private pilot. I am the perfect candidate for the Mars One team because I am a fun, intelligent, independent woman with a creative and resourceful mind.

Interests:

I am interested in Bioastronautics and the advancement of humanity. I enjoy the simple things in life like spending quality time with my family and friends, keeping a healthy mind and body, and being part of programs that promote interest in Science, Technology, Engineering and Mathematics.

Profile 7: Oscar.

Oscar has perhaps the most astronaut-worthy resume. He's a pilot, he's in the military, and he's on his way to checking off his PhD. Good job Oscar. He has a strong set of avocational interests that include scuba-diving and 3D printers. If I was on the Mars One selection committee I would list him as a likely candidate. Having said that, I suggest he save himself and apply to NASA instead.

Name:	Oscar
Sex:	Male
Age:	34
From:	United States
Language:	English

Self Introduction:

I'm Oscar, and I would like to explore the vastness of Mars to search for extant or past signs of life. I am a Valedictorian, an Eagle Scout, an AF Academy grad, and a licensed pilot. I work as an Aerospace Engineer on tactical aircraft for NAVAIR at Pax River NAS, and I previously worked as a practicing Nuclear Test Engineer on Naval reactors. As a Navy Reservist, I am a Flight Test Engineer at NAS Pax River. During my Masters in Aerospace, I completed research at NASA Ames leading to a thesis concerning aeolian dune formation by particle saltation on Saturn's moon, Titan. Recently, I earned candidacy for the Aerospace PhD program at Old Dominion University (ODU) in Norfolk, VA with a focus on spacecraft and habitat radiation shielding. In my spare time, I cycle, SCUBA, and tinker with 3D printers/UAV's. Join me on a Beautiful New World?

Interests:

- Virtual Reality/VR gaming & simulation
- 3D Selective Laser Sintering/Printing
- Camping/Hiking/Wilderness Survival
- Arduino
- Space
- Trumpet
- Soccer/Running
- Flying

- Music
- Reading
- Philosophy

Follow me on Twitter: @Astro_Osk

Profile 8: Lucie.

Lucy or Lucie is a fan of Elon Musk, which is tinged with irony since it will almost certainly be Elon who is the driver behind having the first humans on Mars. Lucie is lightweight in terms of qualifications though. Still, who needs qualifications to get to Mars? Not reality show contestants apparently.

Name:	Lucy
Sex:	Female
Age:	26
From:	Czech Republic
Language:	English

Self Introduction:

Hi, I'm Lucie and I'm the Czech candidate among the Mars One 100. I reside in the UK at the moment and oscillate between my job as a physics technician and independent study of SciTech-related subject areas. I'm bilingual, rational, curious and increasingly worried about the global issues clouding our bright future. I'm on a quest to achieve the levels of discipline, knowledge and skill to be able to meaningfully improve said future, while maintaining the humility, empathy and sense of duty necessary for guiding me on that path.

Interests:

I enjoy working out, writing, spending time outdoors, meeting new people, acquiring new skills, thinking and having witty and/or intellectual conversations. I like hands-on, practical experiences, finding new motivation and inspiration - and passing it on to others. I am a huge fan of Mass Effect, Elon Musk, AI, various cosmological theories and the human potential.

Profile 9: Kellie.

I've met Kellie and she's a bright spark who's the life and soul of any group gathering: the most gregarious person you could ever hope to meet and someone who would brighten up any mission, even one as ill-conceived as Mars One. A public figure who is a tireless advocator for manned spaceflight, Kellie also happens to have the most stellar of all the Mars One resumes. And she's only 27. Mars One selection committee take note!

Name:	Kellie
Sex:	Female
Age:	27
From:	United States
Language:	English

Self Introduction:

Kellie Gerardi is a space science communicator who has worked with a number of commercial space companies, non-profit organizations, and government agencies to research, develop, empower, and communicate the progress of the spaceflight industry. Gerardi currently leads business development efforts at Masten Space Systems, an aerospace R&D and rocket flight services company known for their innovative work with DARPA, NASA, and industry customers. She also serves as the Media Specialist for the Commercial Spaceflight Federation (CSF), the United States' industry trade association of leading commercial companies working to advance human spaceflight.

Gerardi is an active member of The Explorers Club, a global scientific organization whose famed Annual Dinners she has Chaired for the past two years. Gerardi recently carried The Explorers Club Flag on a two-week expedition to the Mars Desert Research Station, a prototype laboratory used by a variety of national space agencies and scientists to conduct analog Martian field research and simulate long-duration spaceflight.

An avid science communicator, Gerardi has authored dozens of essays, white papers, and research studies. She has appeared on popular television shows such as ABC's "The View" and "Nightline" and has been interviewed on multiple NPR podcasts. She has been profiled in a variety of popular publications, including The New York Times, the Huffington Post, Popular Mechanics, Popular Science, and Vogue. Her thought leadership enabled her to become a science and exploration influencer to over 15,000 followers across social-media platforms.

In 2015, Gerardi was named a "Rising Talent" by the International Women's Forum for the Economy and Society, an initiative aiming to distinguish talented young women on their way to becoming influential figures in global economies and societies. She was also competitively selected by the Kruger Cowne Agency as a "Rising Star", currently one of 30 finalists under consideration for a trip to space.

Interests:

Pushing the boundaries of science, technology, and exploration.

Profile 10: Robin.

Robin is from my home country of Norway. He has the right intentions and the right plans and he comes from a nation of some of the toughest explorers – Amundsen, Nansen, Heyerdahl, to name but a few. But while these explorers embarked upon very risky expeditions, they steered away from ventures that had suicide written all over them. Think again Robin.

Name:	Robin
Sex:	Male
Age:	21
From:	Norway
Language:	English

Self Introduction:

Greetings, my name is Robin and I'm 20 years old.

I have been interested in space since I was 4-5 years old and saw the movie "Apollo 13" for the first time. Since then, I have had a dream about exploring space, and joining the colonization of another planet.

I am currently a student taking a few classes over again to improve my grades. When I am done with this I am going to take a bachelor in engineering, or become a pilot. If I become an engineer, I will most likely take a master's degree after the bachelor is done. Interests:

I am in love with space and everything related to space. I also enjoy reading a good book. My favorite authors are Michael Cobley and Wilbur Smith.

I love to fly airplanes, and because of this I have started on my PPL-A (Private Pilot License – Airplane). I am currently reading the theory and booked my first hours.

> "Yes, but to travel to another planet, knowing you can never come back, you'd have to be pretty sad. Aniston sad."
>
> *Marge Simpson, The Simpsons, Season 27 Episode 16.*
> The Marge-ian Chronicles. *Original Air Date, March 13, 2016*

So how do the Mars One 100 applicants stack up against professional astronauts? Well, there are some in the group who might have a good chance of making it to the final 80 to 100 candidates in a regular astronaut selection (two of the Canadian Mars One candidates applied to the Canadian Space Agency's 2016 Astronaut Recruitment Campaign), but as we've discussed in this book, launching to low Earth orbit and travelling to Mars are two completely different balls of wax. But that is precisely the appeal of Mars One: taking supposedly regular people off the street and asking if they have the right stuff.

A similar show aired on British TV in 2015. It was called *SAS: Who Dares Wins*. The idea behind the show was to see what happened when 30 normal men spend eight days completing a shortened version of the selection process used by British Special Forces. The program featured a lot of scenes with instructors sticking bags over people's heads, taking them to interrogation rooms and making them cry. In the regular selection process, of 200 soldiers that begin, only between 5 and 10 percent, make it. And the process of whittling those numbers down is very ugly indeed. Which is probably what will happen in Mars One.

Index